Springer Water

The book series Springer Water comprises a broad portfolio of multi- and interdisciplinary scientific books, aiming at researchers, students, and everyone interested in water-related science. The series includes peer-reviewed monographs, edited volumes, textbooks, and conference proceedings. Its volumes combine all kinds of water-related research areas, such as: the movement, distribution and quality of freshwater; water resources; the quality and pollution of water and its influence on health; the water industry including drinking water, wastewater, and desalination services and technologies; water history; as well as water management and the governmental, political, developmental, and ethical aspects of water.

Gennaro Gentile · Mariacristina Cocca ·
Daniela Capuano · Federico Olivieri ·
Roberto Avolio · Rachele Castaldo ·
Edoardo Bemporad · Maria Emanuela Errico
Editors

Proceedings of the 4th International Conference on Microplastic Pollution in the Mediterranean Sea

Springer

Editors
Gennaro Gentile
Institute of Polymers, Composites and Biomaterials
National Research Council of Italy
Pozzuoli, Naples, Italy

Mariacristina Cocca
Institute of Polymers, Composites and Biomaterials
National Research Council of Italy
Pozzuoli, Naples, Italy

Daniela Capuano
Institute of Polymers, Composites and Biomaterials
National Research Council of Italy
Pozzuoli, Naples, Italy

Federico Olivieri
Institute of Polymers, Composites and Biomaterials
National Research Council of Italy
Pozzuoli, Naples, Italy

Roberto Avolio
Institute of Polymers, Composites and Biomaterials
National Research Council of Italy
Pozzuoli, Naples, Italy

Rachele Castaldo
Institute of Polymers, Composites and Biomaterials
National Research Council of Italy
Pozzuoli, Naples, Italy

Edoardo Bemporad
Institute of Polymers, Composites and Biomaterials
National Research Council of Italy
Pozzuoli, Naples, Italy

Maria Emanuela Errico
Institute of Polymers, Composites and Biomaterials
National Research Council of Italy
Pozzuoli, Naples, Italy

ISSN 2364-6934 ISSN 2364-8198 (electronic)
Springer Water
ISBN 978-3-032-23846-7 ISBN 978-3-032-23847-4 (eBook)
https://doi.org/10.1007/978-3-032-23847-4

This Springer imprint is published by the registered company Springer Nature Switzerland AG
The registered company address is: Gewerbestrasse 11, 6330 Cham, Switzerland

Contents

Chapter 1
From Soil to Sea: Environmental Degradation of PLA–Glass Fibre Composites

Sarai Agustin-Salazar, Lucia Conzatti, Roberto Utzeri, Lorena Affatato, Maria Gelli, Michela Tassistro, Ilaria Schizzi, Francesca Spanò, Pierfrancesco Cerruti, Valentina Marturano, Tiziana Marino, Paola Stagnaro, and Gabriella Santagata

Introduction

Synthetic plastics have profoundly transformed modern society due to their low cost, versatility, and wide range of applications; however, their increasing production from fossil resources has led to serious environmental challenges, including resource depletion, greenhouse gas emissions, pollution, and microplastic accumulation (Kachrimanidou et al. 2021). In response, regulatory actions such as EU Directive 2019/904 have promoted a transition toward more sustainable alternatives, including bio-based and biodegradable plastics (Pellis et al. 2021). Bioplastics, defined as polymers that are bio-based, biodegradable, or both, represent a promising solution, although their terminology and classification can be misleading (Agustin-Salazar et al. 2025). They can be categorized into bio-based non-biodegradable polymers, bio-based biodegradable polymers such as polylactic acid (PLA), and biodegradable fossil-based polymers (Pellis et al. 2021). Despite their advantages—reduced reliance on fossil resources, potential carbon neutrality, and biodegradability—their market share remains limited, mainly due to high production costs, performance

S. Agustin-Salazar (✉) · L. Affatato · M. Gelli · P. Cerruti · V. Marturano · T. Marino · G. Santagata
Institute for Polymers, Composites and Biomaterials (IPCB-CNR), Pozzuoli, Italy
e-mail: sarai.agustinsalazar@cnr.it

L. Conzatti · R. Utzeri · M. Tassistro · I. Schizzi · P. Stagnaro
Institute of Chemical Sciences and Technologies "Giulio Natta" (SCITEC-CNR), Genova, Italy

F. Spanò
Institute of Biophysics (IBF-CNR), Torre di Francia, Genova, Italy

G. Gentile et al. (eds.), *Proceedings of the 4th International Conference on Microplastic Pollution in the Mediterranean Sea*, Springer Water,
https://doi.org/10.1007/978-3-032-23847-4_1

constraints, land and water use concerns, and insufficient waste management and composting infrastructure (Stasiškienė et al. 2022; Agustin-Salazar et al. 2025).

Among bioplastics, PLA is the most widely used thanks to its biocompatibility, processability, and mechanical performance, but its biodegradation strongly depends on environmental conditions and material crystallinity (Chamas et al. 2020; Rezvani Ghomi et al. 2021). Reinforcement strategies, including the use of innovative degradable glass fibers, are being developed to improve PLA performance while preserving sustainability (Sun et al. 2022). In this context, ongoing academic and industrial research, including the EU-funded BIOPLASTICS EUROPE project, aims to develop high-performance, sustainable PLA-based materials and to evaluate their degradation behavior in soil and marine environments for short- to medium-term applications. In this study, a PLA-based composite (PLA-DGF) containing 20 wt.% degradable glass fiber ArcBiox™ (https://abmcomposite.com/degradable-glass-fibers/) was subjected to long-term degradation in Mediterranean seawater (MS), soil burial (DS), and accelerated photo-oxidation (POx) in a climatic chamber (Agustin-Salazar et al. 2025).

Materials and Methods

On-field degradation tests were conducted on PLA-DGF dumbbell specimens (geometry 1A) exposed to circulating Mediterranean seawater in outdoor tanks starting from mid-October 2020. Environmental parameters were monitored over one year, with temperature ranging from 11 to 26 °C, pH between 8.1 and 8.4, salinity from 36.3 to 42.9 g kg^{-1}, and dissolved oxygen between 44 and 51%. Samples were characterized in the pristine state and after 1, 3, 6, and 12 months of exposure to evaluate changes in thermal and mechanical properties.

PLA-DGF films were obtained by compression molding from dumbbell bars and shaped into dog-bone specimens for accelerated ageing and soil degradation tests. Accelerated ageing was carried out in a climatic chamber equipped with UV irradiation ($\lambda > 250$ nm) at 40 °C and 50% relative humidity, with analyses performed up to 1036 h of exposure. Soil degradation tests were conducted by burying conditioned specimens at a depth of 5 cm in commercial soil under controlled moisture conditions, monitoring material evolution up to 1008 h.

Thermal properties were analyzed by thermogravimetric analysis (TGA) and differential scanning calorimetry (DSC), determining degradation temperatures and thermal transitions. Mechanical performance was evaluated by uniaxial tensile tests on both dumbbell bars and film specimens according to relevant ISO standards, allowing assessment of the effects of environmental exposure on material integrity.

Results and Discussion

Test bars were exposed to Mediterranean seawater for up to 12 months and analysed by DSC, TGA, and tensile tests. Thermal properties showed minimal variation over time, with only a slight decrease in thermal stability. DSC analysis revealed limited changes. Crystallinity decreased from 54 to 38% after 12 months. Mechanical behaviour remained brittle, with a gradual decay of properties after 12 months (Fig. 1.1). PLA-DGF films ($\geq$ 150 μm) were obtained from the bars and subjected to accelerated degradation in soil (DS, 1008 h) and photo-oxidation (POx, 1036 h). In Table 1.1 a comparison among the studied environments is presented.

Optical images showed surface opacity and erosion after DS, while POx induced rapid yellowing within 60 h. TGA analysis evidenced a reduction of T_{onset} only after POx, whereas DS did not significantly affect thermal stability. The two-step degradation behavior was attributed to PLA and PBAT fractions (Xiang et al. 2020; Agustin-Salazar et al. 2022a; b).

DSC analysis confirmed the presence of PLA and PBAT thermal transitions (Han et al. 2020), with PLA crystallinity slightly increasing after film processing (Palsikowski et al. 2018). After DS, dual melting endotherms associated with α' and α crystalline forms were observed (Cocca et al. 2011; Di Lorenzo and Androsch 2019; Agustin-Salazar et al. 2022a). POx had a stronger effect, causing reductions in melting temperatures and enthalpies, whereas soil ageing induced negligible thermal changes. Mechanical properties decreased by about 50% after film processing. Soil degradation led to progressive embrittlement, preventing tensile testing after six weeks, while POx caused faster degradation, particularly in strain (− 52% after 160 h), followed by stress and modulus reductions (Fig. 1.1).

After 12 months in seawater, no significant mass loss or visual damage was observed, apart from surface fouling. TGA and ATR-FTIR spectra remained unchanged, while DSC showed a ~ 30% decrease in PLA crystallinity and increased melting temperatures. Tensile tests confirmed mechanical decay, with a 55% reduction in elastic modulus and halved breaking force (Urbanek et al. 2021; Salazar et al. 2023).

Conclusions

PLA degradation in Mediterranean seawater was extremely slow, with detectable changes in crystallinity only after 12 months of exposure. In contrast, PLA-based composites containing degradable glass fibers showed significant susceptibility to soil degradation and photooxidation, with UV ageing markedly affecting thermal properties. These results confirm that PLA degradation is highly dependent on both material formulation and environmental conditions, emphasizing the need to consider end-use scenarios when designing sustainable PLA-based products.

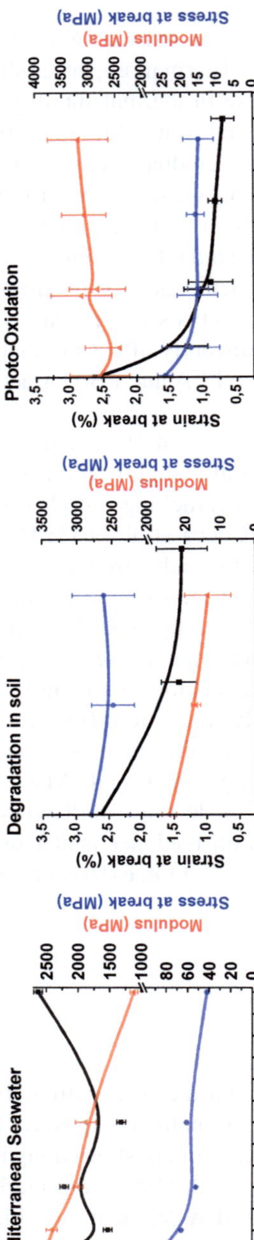

Fig. 1.1 Changes in mechanical properties during degradation, MS, DS and POx, of PLA-DGF films

Table 1.1 PLA-DGF degradation in different environments

	Mediterranean seawater	Degradation in soil	Photo-oxidation
Visual	Stable, white surface	Opaque, slight erosion	Yellowing, opacity ↑
Thermal	Minor stability loss; χc ↓ (54 → 38%)	Unchanged	Early degradation; enthalpy ↓
Mechanical	Modulus and stress ↓ ~ 50%; brittle	Strength ↓ ~ 50% (336 h); untestable after 6 w	Rapid decay; strain − 52% (160 h)
Rate	Slow	Moderate	Fast
Mechanism	Hydrolysis	Biodegradation	Photo-oxidation

Acknowledgments The authors gratefully acknowledge the support of the projects "U H2020 CE-BG-06-2019 BIO-PLASTICS EUROPE", "PET2POLY (CUP B53D23015850006)", HE CBE n 101130442. Biodegradable delivery systems for plant pathogens control of horticultural crops through "BIoactiVE agents" Acron BioBIVE (CUP B87F24000050006). HE CBE n. 101156998. "Recycling locally produced bio-wastes to ensure affordability and availability of innovative bio-based fertilisers" Acron. ReLeaf (CUP B87F24000030006). This paper is dedicated to the memory of Mario Malinconico.

References

Agustin-Salazar S, Conzatti L, Utzeri R, Affatato L, Gelli M, Tassistro M, Schizzi I, Spanò F, Cerruti P, Stagnaro P, Santagata G (2025) Bio-based plastics: challenge or reality? Perspectives and constraints of their exploitation as sustainable alternative to fossil plastics, pp 387–406

Agustin-Salazar S, Ricciulli M, Ambrogi V, Cerruti P, Scarinzi G (2022a) Effect of thermal annealing and filler ball-milling on the properties of highly filled polylactic acid/pecan nutshell biocomposites. Int J Biol Macromol 200:350–361. https://doi.org/10.1016/j.ijbiomac.2021.12.101

Agustin-Salazar S, Ricciulli M, Ambrogi V, Cerruti P, Scarinzi G (2022b) Thermomechanical properties and biodegradation behavior of itaconic anhydride-grafted PLA/pecan nutshell biocomposites †. Polymers (Basel) 14(24):1–20. https://doi.org/10.3390/polym14245532

Chamas A, Moon H, Zheng J, Qiu Y, Tabassum T, Jang JH, Abu-Omar M, Scott SL, Suh S (2020) Degradation rates of plastics in the environment. ACS Sustain Chem Eng 8(9):3494–3511. https://doi.org/10.1021/acssuschemeng.9b06635

Cocca M, Di LML, Malinconico M, Frezza V (2011) Influence of crystal polymorphism on mechanical and barrier properties of poly(l-lactic acid). Eur Polym J 47(5):1073–1080. https://doi.org/10.1016/j.eurpolymj.2011.02.009

Di Lorenzo ML, Androsch R (2019) Thermal properties of bio-based polymers, 1st edn. Springer Nature Switzerland AG, Gewerbestrasse, Cham

Han Y, Shi J, Mao L, Wang Z, Zhang L (2020) Improvement of compatibility and mechanical performances of PLA/PBAT composites with epoxidized soybean oil as compatibilizer. Ind Eng Chem Res 59(50):21779–21790. https://doi.org/10.1021/acs.iecr.0c04285

Kachrimanidou V, Ioannidou SM, Ladakis D, Papapostolou H, Kopsahelis N, Koutinas AA, Kookos IK (2021) Techno-economic evaluation and life-cycle assessment of poly(3-hydroxybutyrate) production within a biorefinery concept using sunflower-based biodiesel industry by-products. Bioresour Technol 326(Oct 2020):124711. https://doi.org/10.1016/j.biortech.2021.124711

Palsikowski PA, Kuchnier CN, Pinheiro IF, Morales AR (2018) Biodegradation in soil of PLA/PBAT blends compatibilized with chain extender. J Polym Environ 26(1):330–341. https://doi.org/10.1007/s10924-017-0951-3

Pellis A, Malinconico M, Guarneri A, Gardossi L (2021) Renewable polymers and plastics: Performance beyond the green. N Biotechnol 60:146–158. https://doi.org/10.1016/j.nbt.2020.10.003

Rezvani Ghomi ER, Khosravi F, Saedi Ardahaei AS, Dai Y, Neisiany RE, Foroughi F, Wu M, Das O, Ramakrishna S (2021) The life cycle assessment for Polylactic Acid (PLA) to make it a low-carbon material. Polymers (Basel) 13(11):1854. https://doi.org/10.3390/polym13111854

Salazar SA, Abdulhameed S, Sánchez M d RS (2023) Biodegradation of polymers. In: Salazar M d R, Solanilla Duque JF, Saenz-Galindo A, Rodriguez-Herrera R (eds) Biodegradable polymers, 1st edn. CRC Press, Boca Raton, pp 1–12

Stasiškienė Ž, Barbir J, Draudvilienė L, Chong ZK, Kuchta K, Voronova V, Leal Filho W (2022) Challenges and strategies for bio-based and biodegradable plastic waste management in Europe. Sustainability 14(24):16476. https://doi.org/10.3390/su142416476

Sun Y, Zheng Z, Wang Y, Yang B, Wang J, Mu W (2022) PLA composites reinforced with rice residues or glass fiber—a review of mechanical properties, thermal properties, and biodegradation properties. Journal of Polymer Research 29

Urbanek AK, Strzelecki MC, Mirończuk AM (2021) The potential of cold-adapted microorganisms for biodegradation of bioplastics. Waste Manag 119:72–81. https://doi.org/10.1016/j.wasman.2020.09.031

Xiang S, Feng L, Bian X, Li G, Chen X (2020) Evaluation of PLA content in PLA/PBAT blends using TGA. Polym Test 81:106211. https://doi.org/10.1016/j.polymertesting.2019.106211

Chapter 2
Navigating the Shift from Fossil-Based Plastics to Bio-Based Alternatives in Maritime Applications

Wajid Ali, Périne Doyen, and Rachid Amara

Introduction

The maritime sector plays a significant role in global socio-economic development, generating trillions of dollars in gross value added and employing millions of people worldwide (OECD 2025). However, the extensive use of plastics in this sector raises serious environmental concerns, as plastics are both released during operations and pose significant challenges at the end of their life. They are widely used in fishing and aquaculture gear, vessel coatings and composites, seafood packaging, and tourism-related activities. The widespread production of conventional plastics poses multiple challenges. Plastics already account for up to 9% of global oil and gas demand and could reach 20% by 2050 (Agenda 2016; Nielsen et al. 2020), with associated emissions of 56 gigatons of CO_2-equivalent (CIEL 2019). Their persistence in the marine environment is a global concern, with the maritime sector contributing up to 20% of total marine plastic inputs. Plastic pollution poses significant ecological risks through entanglement, ingestion, habitat alteration, and leaching of hazardous additives. To mitigate these impacts, bio-based plastics have been promoted as potential alternatives (Ali et al. 2023), with applications in the maritime sector now emerging. This chapter provides an overview of these applications, focusing on their environmental, economic, and social implications, along with commonly used analytical methods for assessment.

W. Ali · P. Doyen
Univ. Littoral Côte d'Opale, UMRt 1158 BioEcoAgro, USC ANSES, INRAe, Univ. Artois, Univ. Lille, Univ. Picardie Jules Verne, Univ. Liège, Junia, Boulogne-sur-Mer, France

R. Amara (✉)
Univ. Littoral Côte d'Opale, CNRS, IRD, Univ. Lille, UMR 8187 - LOG – Laboratoire d'Océanologie et de Géosciences, Wimereux, France
e-mail: rachid.amara@univ-littoral.fr

G. Gentile et al. (eds.), *Proceedings of the 4th International Conference on Microplastic Pollution in the Mediterranean Sea*, Springer Water,
https://doi.org/10.1007/978-3-032-23847-4_2

Biobased Plastics in Maritime Applications

Among bio-based materials, bio-based plastics have attracted particular attention for their potential to replace conventional plastics. Derived from renewable feedstocks, these materials can be tailored to specific applications. In the maritime sector, adoption remains at an early stage, with several pilot-scale initiatives underway. Current efforts include the design of biodegradable fishing gear, the development of bio-based composites for vessels and infrastructure, and sustainable packaging for seafood and tourism activities.

Fisheries and Aquaculture

Pilot-scale and field trials have tested the feasibility of bio-based plastics in fisheries and aquaculture, though technical challenges persist. For example, polylactic acid (PLA) monofilament gillnets generally exhibited weaker physical properties than polyamide (PA) (Shu et al. 2021), although no significant differences in catch performance were observed in a commercial mullet fishery (Yu et al. 2023). In aquaculture, PLA droplines and natural fibers have been tested as alternatives to polypropylene (PP), with mixed results regarding mechanical performance and durability (Goedefroo et al. 2022). The development of several novel biodegradable prototypes further demonstrates the diversity of approaches under exploration, while highlighting the ongoing challenge of balancing functionality, durability, and performance.

Bio-Based Food Packaging in the Seafood and Tourism Sectors

Packaging is the largest application of bioplastics, accounting for about 45% of the global market in 2024, with increasing use in seafood packaging. PLA is the most widely applied polymer, particularly in active packaging systems designed to improve microbial stability, inhibit oxidation, and extend product shelf life (Sourav et al. 2025). In parallel, natural biopolymers such as starch, gelatin, and alginate are being developed for films and trays, showing promising performance in seafood storage (Laorenza et al. 2022). These advances are supported by systemic initiatives, such as the Global Tourism Plastics Initiative, which promotes circular approaches and encourages the adoption of reusable, recyclable, or compostable alternatives across tourism and food distribution chains (UNEP 2020). Together, these efforts highlight the potential of bio-based packaging to reduce environmental impacts while preserving seafood quality and safety.

Maritime Structures

Conventional composites used in shipbuilding and offshore structures are associated with high energy demands and substantial carbon footprints, which has driven interest in bio-composites as more sustainable alternatives, although large-scale production remains limited. Trials with bio-composites have demonstrated promising structural performance, including reduced weight and improved material compatibility. In shipbuilding, prototypes such as hybrid jute or flax-based boats, fully bio-composite vessels, racing yachts, surfboards, and inflatable boats illustrate ongoing efforts to advance bio-based materials (Phiri et al. 2023). These initiatives demonstrate progress toward integrating bio-composites into maritime applications and aligning structural design with sustainability goals.

Environmental and Socio-Economic Aspects

The environmental benefits of bio-based plastics are often linked to their renewable origins. However, the use of first-generation feedstocks can drive indirect land-use change, biodiversity loss, and fertilizer-related emissions. Second-generation feedstocks provide more sustainable options with lower competition for arable land but remain constrained by technological limitations (Singh et al. 2022). Climate benefits are highly context-dependent, varying with feedstock type, polymer pathway, energy source, and life cycle assessment (LCA) methods. End-of-life management is largely similar to that of conventional plastics, except that some bioplastics are compostable (Ali et al. 2023). In marine environments, however, they may still fragment, posing ecotoxicological risks (Ali et al. 2024).

The adoption of bio-based plastics is shaped by both economic feasibility and social acceptance. Their higher costs stem largely from feedstock and chemical inputs, which account for nearly half of total manufacturing expenses (Abrha et al. 2022). Techno-economic assessments (TEA) indicate that competitiveness depends on feedstock type, process configuration, and infrastructure, with promising pathways involving low-cost resources. Social acceptance is likewise influenced by product cost, durability, and environmental awareness. In fisheries, conventional plastic nets are often replaced due to their low price, while bio-based alternatives may face resistance owing to higher costs and weaker mechanical properties, potentially increasing gear loss and replacement frequency. However, evidence from the English Channel suggests that technical performance, rather than cost alone, remains the main barrier to adoption (Drakeford et al. 2023).

Analytical Assessment Methods

A range of analytical frameworks is used to assess the sustainability of bioplastics, with TEA and LCA being the most widely applied. TEA frequently highlights constraints such as high feedstock costs, limited production scales, and emerging technologies, although the use of secondary feedstocks such as biogas and wastewater can improve cost-effectiveness and resource efficiency (Davidson et al. 2021). In parallel, LCA is applied to compare bioplastics with fossil-based plastics, but outcomes vary depending on system boundaries, feedstock type, and selected impact categories, complicating generalizations (Walker and Rothman 2020). Degradation is largely restricted to microbial activity, overlooking ecological processes such as bioaccumulation, biomagnification, and biotransformation, which remain difficult to integrate into LCA (Ali et al. 2023). Overall, TEA and LCA provide valuable insights into the economic and environmental performance of bioplastics, yet both face methodological limitations.

Conclusions

Bio-based plastics are emerging as alternatives to fossil-based plastics. Their application in marine sectors remains limited to trials, constrained by challenges in performance, durability, biodegradability under natural conditions, and microplastic formation. Environmental sustainability depends on feedstock origin, land use, energy inputs, and end-of-life options, while socio-economic barriers include high costs, limited infrastructure, and uncertain acceptance. Sustainable deployment requires environmentally benign formulations with controlled degradation and robust evaluation. Refined TEA and LCA are essential to guide innovation. Future efforts should focus on high-performance, cost-effective, and sustainable bioplastics tailored to marine applications.

References

Abrha H, Cabrera J, Dai Y, Irfan M, Toma A, Jiao S, Liu X (2022) Bio-based plastics production, impact and end of life: a literature review and content analysis. Sustainability 14(8):4855

Agenda I (2016) The new plastics economy: rethinking the future of plastics. In: World economic forum, January, vol 36

Ali W, Ali H, Souissi S, Zinck P (2023) Are bioplastics an ecofriendly alternative to fossil fuel plastics? Environ Chem Lett 21(4):1991–2002

Ali W, Jeong H, Lee JS, Zinck P, Souissi S (2024) Biodegradable microplastics interaction with pollutants and their potential toxicity for aquatic biota: a review. Environ Chem Lett 22(3):1185–1220

CIEL (2019) Plastic & climate: the hidden costs of a plastic planet. CIEL

Davidson MG, Elgie S, Parsons S, Young TJ (2021) Production of HMF, FDCA and their derived products: a review of life cycle assessment (LCA) and tech-no-economic analysis (TEA) studies. Green Chem 23(9):3154–3171

Drakeford BM, Forse A, Failler P (2023) The economic impacts of introducing biodegradable fishing gear as a ghost fishing mitigation in the English Channel static gear fishery. Mar Pollut Bull 192:114918

Goedefroo N, Benham P, Debusschere E, Deneudt K, Mascart T, Semeraro A, Sterckx T, Van Hoey G (2022) Nature-based solutions in a sandy foreshore: a biological assessment of a longline mussel aquaculture technique to establish subtidal reefs. Ecol Eng 185:106807

Laorenza Y, Chonhenchob V, Bumbudsanpharoke N, Jittanit W, Sae-Tan S, Rachtanapun C, Chanput WP, Charoensiddhi S, Srisa A, Promhuad K, Wongphan P (2022) Polymeric packaging applications for seafood products: packaging-deterioration relevance, technology and trends. Polymers 14(18):3706

Nielsen TD, Hasselbalch J, Holmberg K, Stripple J (2020) Politics and the plastic crisis: a review throughout the plastic life cycle. Wiley Interdiscip Rev Energy Environ 9(1):e360

OECD (2025) OECD urges strengthened co-operation to sustain trillion-dollar ocean economy. Press Release, 31 Mar

Phiri R, Rangappa SM, Siengchin S, Oladijo OP, Dhakal HN (2023) Development of sustainable biopolymer-based composites for lightweight applications from agricultural waste biomass: a review. Adv Ind Eng Polym Res 6(4):436–450

Shu AY, Zhang M, Yu WW, Wang Y, Shi JG, Wang L, Min MH (2021) Comparative analysis of physical performance and fishing efficiency between biodegradable PLA gill net and conventional PA gill net

Singh N, Ogunseitan OA, Wong MH, Tang Y (2022) Sustainable materials alternative to petrochemical plastics pollution: a review analysis. Sustain HorizS 2:100016

Sourav C, Sasidharan A, Mohan C, Sabu S, Sunooj KV (2025) Prospects of biodegradable packaging in the seafood industry: a review. Asian Fish Sci 38:50–65

UNEP (2020) Tourism to tackle plastic pollution with new commitment. United Nations Environment Programme

Walker S, Rothman R (2020) Life cycle assessment of bio-based and fossil-based plastic: a review. J Clean Prod 261:121158

Yu M, Tang Y, Min M, Herrmann B, Cerbule K, Liu C, Dou Y, Zhang L (2023) Comparison of physical properties and fishing performance between biodegradable PLA and conventional PA trammel nets in grey mullet (Mugil cephalus) and red-lip mullet (Liza haematocheila) fishery. Mar Pollut Bull 195:115545

Chapter 3
Microplastics in Mussels (*Mytilus galloprovincialis*): A Nationwide Survey of Marine Pollution in Italian Coastal Waters

Erika Baldini, Silva Rubini, Martina Munari, Filippo Barsi, Daniela Meloni, Nicola Pussini, Francesca Barchiesi, Gabriella Di Francesco, Carmen Losasso, Cristiano Cocumelli, Salvatore Dara, Sebastiano Virgilio, Fabio Di Nocera, Antonio Petrella, Matteo Zinni, Carmela Vaccaro, Negar Eftekhari, Stefano Manfredini, and Silvia Vertuani

E. Baldini (✉) · S. Manfredini · S. Vertuani
Department of Life Sciences and Biotechnology, Faculty of Medicine, Pharmacy and Prevention, University of Ferrara, Ferrara, Italy
e-mail: bldrke@unife.it

S. Rubini · M. Munari · F. Barsi
Istituto Zooprofilattico Sperimentale della Lombardia e dell'Emilia Romagna, Ferrara Territorial Office, Ferrara, Italy

D. Meloni
Istituto Zooprofilattico Sperimentale del Piemonte, Liguria e Valle d'Aosta, Torino, Italy

N. Pussini
Istituto Zooprofilattico Sperimentale del Piemonte, Liguria e Valle d'Aosta; Savona – Imperia Territorial Office, Savona, Italy

F. Barchiesi
Istituto Zooprofilattico Sperimentale dell'Umbria e delle Marche, Ancona Territorial Office, Ancona, Italy

G. Di Francesco
Istituto Zooprofilattico Sperimentale dell'Abruzzo e del Molise, Teramo, Italy

C. Losasso
Istituto Zooprofilattico Sperimentale delle Venezie, Legnaro, Italy

C. Cocumelli
Istituto Zooprofilattico Sperimentale del Lazio e della Toscana, Rome, Italy

G. Gentile et al. (eds.), *Proceedings of the 4th International Conference on Microplastic Pollution in the Mediterranean Sea*, Springer Water,
https://doi.org/10.1007/978-3-032-23847-4_3

Introduction

Marine litter is now recognized as one of the greatest environmental challenges of our century, with plastics accounting for almost 80% of the debris found in oceans and seas (Martellini et al. 2018). Their persistence leads to the formation of microplastics (MPs), defined as particles ranging in size from 0.1 μm to 5 mm. Due to their small size and ubiquity, MPs can be ingested by marine organisms and have the potential to reach humans through seafood consumption.

The Mediterranean Sea is recognized as one of the most contaminated areas by MPs due to its semi-enclosed geography, dense coastal population, industrial activities, and riverine inputs (Martellini et al. 2018). In particular, the northern Adriatic is heavily affected by discharges from the Po River, one of Europe's main sources of agricultural, industrial, and urban waste.

Mussels, particularly *Mytilus galloprovincialis*, are valuable bioindicators of marine pollution. As sedentary filter feeders, they easily accumulate microplastics, and since mussels are consumed whole, including their digestive tract, the microplastics they accumulate enter the human diet directly. This creates a direct pathway for human exposure to MPs and the pollutants they may carry. Indeed, MPs can act as carriers of persistent organic pollutants (POPs), heavy metals, and pathogenic microorganisms, raising significant concerns about the safety of seafood products and long-term health risks (Bouwmeester et al. 2015).

This study presents the first nationwide survey of MPs contamination in mussels from Italian coastal waters, providing a comprehensive picture of regional contamination and its implications for both ecosystem health and food safety.

S. Dara
Istituto Zooprofilattico Sperimentale della Sicilia, Palermo, Italy

S. Virgilio
Istituto Zooprofilattico Sperimentale della Sardegna, Sassari, Italy

F. Di Nocera
Istituto Zooprofilattico Sperimentale del Mezzogiorno, Portici, Italy

A. Petrella
Istituto Zooprofilattico Sperimentale della Puglia e della Basilicata, Foggia, Italy

M. Zinni
Microbiologia Ambientale e Molecolare – MICAMO Lab Università degli Studi di Genova, Genoa, Italy

C. Vaccaro · N. Eftekhari
Department of Environmental and Prevention Sciences, University of Ferrara, Ferrara, Italy

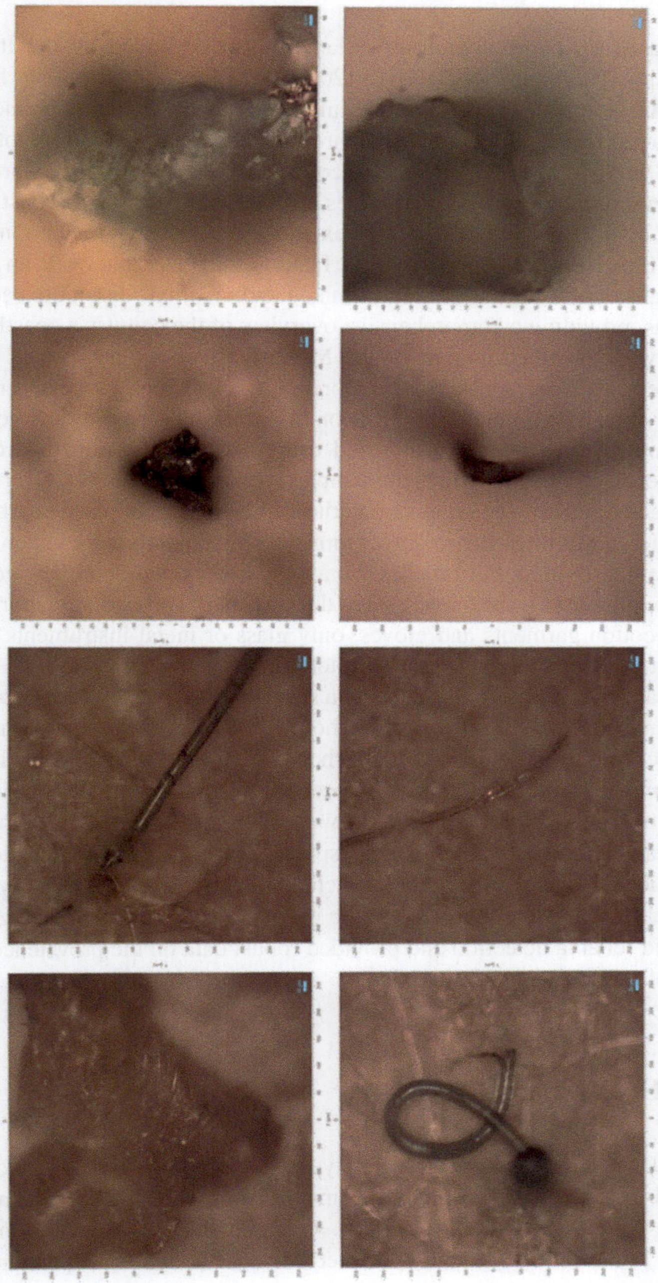

Fig. 3.1 Representative microplastic particles identified in *Mytilus galloprovincialis* samples, including filaments and fragments of various polymers (polystyrene, PVC, polypropylene) and colors (transparent, blue, green). Particle sizes ranged from 32 to 581 µm

Materials and Methods

This study is based on two national surveillance projects funded by the Italian Ministry of Health. The first, PRC2019015 (2020–2023), examined mussel farms in the northern Adriatic Sea, an area strongly influenced by the Po River plume and therefore considered highly vulnerable to anthropogenic contamination. A total of 146 samples of *Mytilus galloprovincialis* were collected from georeferenced farming sites located between three to six nautical miles off the Emilia-Romagna coast. The second, PRC2021101 Strategic (2021–2024), extended the investigation nationwide and involved ten Experimental Zooprophylactic Institutes. In this framework, an additional 104 mussel samples were collected from aquaculture areas distributed across 11 Italian regions, providing a comprehensive overview of the country's coastline.

All samples were obtained as part of the Mollusk Bivalve Surveillance Plan. Immediately after collection, mussels were stored at − 20 °C and processed in the laboratory under strict contamination control conditions. Soft tissues were dissected, pooled into composite samples, and digested using a 10% potassium hydroxide solution (Thiele et al. 2019), after which the samples were incubated at 60 °C for 48 h. This method effectively removed organic material while preserving plastic particles. The digested material was filtered through 0.1 mm glass fiber membranes, which were then dried and stored in glass Petri dishes. To avoid cross-contamination, rigorous quality assurance measures were implemented throughout the procedure: laboratory personnel wore cotton garments and gloves, only glass or metal instruments were used, and all water was deionized and microfiltered.

Microscopic analysis was carried out using a stereomicroscope to identify potential MPs based on their morphology, color, and homogeneity. Criteria for preliminary identification included a particle size of less than 5 mm, uniform thickness in the case of fibers, uniform coloration, and the absence of organic matter adhering to the surface. Particles fulfilling these criteria were then subjected to confirmation and chemical characterization using μ-Raman spectroscopy. This technique allowed for the precise identification of polymers, including polypropylene, polystyrene, and PVC, as well as the detection of synthetic dyes and pigments. This integrated approach ensured reliable detection and provided robust data on the prevalence and characteristics of microplastics in mussels.

Results and Discussion

A total of 250 mussel samples were analyzed, combining the two projects PRC2019015 (146 samples from Emilia-Romagna) and PRC2021101 Strategic (104 samples from different Italian regions). Overall, 7% of samples from the first project and 13% from the second contained MPs. Although these percentages may appear relatively low, they are consistent with international studies and highlight the widespread but heterogeneous presence of MPs in Mediterranean coastal environments.

Regional Prevalence

Marked geographical variability in the prevalence of MPs was observed. Mussels from the northern Adriatic, influenced by the Po River plume, had higher contamination levels than those from southern and insular regions. Data from the PRC2021101 Strategic project show differences among the institutes involved, highlighting the influence of local pollution sources and hydrodynamic conditions on MPs distribution. These results underscore the role of mussels as effective sentinels of localized contamination and provide valuable reference data for regional monitoring.

Morphology and Particle Characteristics

The MPs identified ranged in size from 32 to 581 μm and were mainly in the form of fragments and fibers. Representative examples are shown in Fig. 3.1, where different morphologies and colors are evident. Most of the particles are fragments (64–70%), while the remaining 30–36% are fibers. Color analysis confirmed the predominance of green (50–60%) and blue (30–36%) particles, with smaller proportions of transparent and black particles.

Raman spectroscopy allowed precise identification of polymer types. The main polymers detected were polypropylene (PP) and polystyrene (PS), each representing 33% of the analyzed samples, along with polyvinyl chloride (PVC) and other less common polymers such as P-ether-sulfone and P-ethylene-co-acryloacid. The identification of these polymers is consistent with their widespread use in packaging, fishing equipment, and industrial processes, confirming their likely anthropogenic origin (Andrady 2011).

Pigments and Dyes

In addition to polymers, μ-Raman analysis also revealed the presence of synthetic dyes, including Phthalocyanine, Pigmosol green, Irgalith blue, and Hostasol green. These pigments are widely used in plastic manufacturing and provide further evidence of industrial origins of MPs.

Conclusions

This nationwide investigation confirms the presence of microplastics in mussels harvested along the Italian coastline. Although the overall prevalence ranged from 7 to 13%, levels varied considerably across regions, with higher contamination levels recorded in areas most influenced by anthropogenic pressures, such as the Po River plume or areas with dense maritime traffic. These findings highlight the role of mussels as sentinels of local environmental quality and provide important baseline data for future monitoring.

From a human health perspective, the detection of microplastics in mussels is of particular concern because these bivalves are consumed whole, including the digestive tract where the particles accumulate. As microplastics can adsorb and transport pollutants such as heavy metals, persistent organic contaminants, and microbial biofilms, their ingestion represents a potential risk to food safety and public health. Although toxicological evidence related to human exposure is still limited, the data presented here underscore the need for precautionary measures.

The study also emphasizes the urgent need for standardized monitoring protocols and policies to reduce plastic emissions, alongside public awareness and interdisciplinary collaboration. Overall, this work provides an important national baseline and highlights the need for further research on the ecological and health impacts of microplastics, as well as on strategies to mitigate their presence in marine ecosystems.

References

Andrady AL (2011) Microplastics in the marine environment. Mar Pollut Bull 62:1596–1605

Bouwmeester H, Hollman PC, Peters RJ (2015) Potential health impact of environmentally released micro- and nanoplastics in the human food production chain: experiences from nanotoxicology. Environ Sci Technol 49:8932–8947

Martellini T, Guerranti C, Scopetani C, Ugolini A, Chelazzi D, Cincinelli A (2018) A snapshot of microplastics in the coastal areas of the Mediterranean Sea. TrAC Trends Anal Chem 109:173–179

Thiele CJ, Hudson MD, Russell AE (2019) Evaluation of existing methods to extract microplastics from bivalve tissue: adapted KOH digestion protocol improves filtration at single-digit pore size. Mar Pollut Bull 142:384–393

Chapter 4 Breathing Plastic: Quantifying Indoor Airborne Microplastics in Industrial Workplaces Using Py–GC–MS and Fluorescence Microscopy

Leonardo Barlucchi, Jacopo La Nasa, Claudia Campanale, Augusta Piccardi, Marcello De Rossi, Francesca Modugno, and Lorenzo Palumbo

Introduction

Microplastics (MPs) have increasingly been recognized as emerging contaminants of environmental and human health concern. MPs are described as plastic particles with dimensions between 1 μm and 1 mm (ISO 2023). MPs can be divided into primary, which are intentionally developed with these dimensions, and secondary, generated through the environmental degradation and fragmentation of larger plastic items, driven by physical, chemical, and biological processes (Song et al. 2024). MPs have been detected in all environmental compartments: sea (Muñiz and Rahman 2025), oceans (Zhao et al. 2025), soil (Sajjad et al. 2022), sediments (Yin 2023), atmosphere (Amato-Lourenço et al. 2022), organisms (Jeong et al. 2024), and humans (Saha and Saha 2024). In recent years, MPs have drawn attention from the scientific community as they can be ingested or inhaled by organisms and humans, potentially leading to adverse health effects, such as oxidative stress, cytotoxicity, distant tissue damage, disruption of immune function, and neurotoxicity (Sadique et al. 2025). The atmosphere compartment can be divided into outdoor and indoor environments (Jahanzaib et al. 2025). Sources of MPs in the atmosphere are mostly the releases

L. Barlucchi (✉) · J. La Nasa · F. Modugno
Department of Chemistry and Industrial Chemistry, University of Pisa, Pisa, Italy
e-mail: leonardo.barlucchi@phd.unipi.it

C. Campanale · A. Piccardi · L. Palumbo
Department of Environmental and Health, National Institute of Health (ISS), Roma, Italy

M. De Rossi
Laboratory of Industrial Hygiene, ASL Viterbo, Viterbo, Italy

G. Gentile et al. (eds.), *Proceedings of the 4th International Conference on Microplastic Pollution in the Mediterranean Sea*, Springer Water,
https://doi.org/10.1007/978-3-032-23847-4_4

of microfibers from textiles, industrial emissions, and the degradation of synthetic polymers, as well as the degradation of tires (O'Brien et al. 2023). MPs concentrations are generally higher in indoor environments than in outdoor environments, likely due to the more confined spaces, reduced ventilation and air exchange, and the presence of specific pollution sources such as clothing and household furnishings (Jahanzaib et al. 2025). Since humans spend most of their time indoors, it is crucial to assess the concentrations of this pollutant, as inhalation of MPs (Yakovenko et al. 2025). Despite the well-documented presence and effects of MPs on the environment, their identification and quantification remain challenging. Visual techniques, such as optical microscopy, represent the easiest way to detect MPs. However, the accuracy of visual analysis in identifying MPs is relatively low, and it is not suitable for MPs with dimensions lower than 50 μm (Kalaronis et al. 2022). Commonly used chemical characterization and analysis techniques for MPs include Fourier Transform Infrared spectroscopy (FTIR) and Raman spectroscopy, thermal analysis, and mass spectrometry. FTIR and Raman spectroscopy are non-destructive techniques that allow the identification of MPs based on their spectral fingerprints. However, FTIR spectroscopy has some limitations, including the presence of water during the analysis and its relatively high detection limit (20 μm) (Käppler et al. 2016). Raman spectroscopy offers several advantages over FTIR, such as broader spectral coverage, lower interference from water, and narrower spectral bands (Käppler et al. 2016). However, these techniques are laborious and time-consuming. To overcome these issues, staining MPs with fluorescent dyes has been proposed as a method to improve these issues. Imaging analysis of MPs with fluorescent dyes is being more widely used due to its ease of application. Nile Red, 9-(diethylamino)-5*H*-benzo[*a*]phenoxazin-5-one, was suggested by Andrady (2011) as a promising dye for enhancing the visual detection and quantification of MPs. Thermal analysis and mass spectrometry techniques, such as pyrolysis-gas chromatography-mass spectrometry (Py–GC–MS), have proven effective for identifying and quantifying MPs (Jiang et al. 2025). During pyrolysis, polymers generate characteristic pyrolysis products that are specific to each polymer type, allowing their identification and quantification (Matsueda et al. 2021). In this study, the main objective is to quantify airborne MPs in indoor working environments, where workers may experience higher exposure levels compared to outdoors. Fluorescence microscopy and Py–GC–MS were used as techniques to assess the concentrations of airborne MPs. By combining these techniques, it is possible to obtain a more comprehensive evaluation of MPs contamination, expressed not only in terms of plastic particle number but also as mass per cubic meter, which is essential for exposure assessment and risk evaluation.

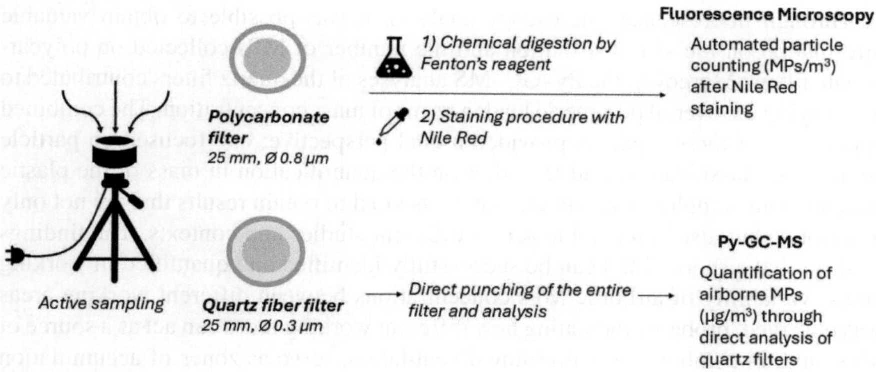

Fig. 4.1 Schematic overview of the analytical workflow applied for the detection, characterization, and quantification of airborne MPs by fluorescence microscopy and Py–GC–MS

Materials and Methods

For this study, the aim was to assess the presence and quantify airborne MPs in indoor working environments. Samples were collected using active samplers in all the working environments. The analysis workflow is schematized in Fig. 4.1. For fluorescence microscopy analysis, the airborne MPs were sampled using polycarbonate filters. Organic material in the samples was digested using standardized digestion procedures, and MPs were subsequently stained using Nile Red fluorescent dye. For Py–GC–MS analysis, the airborne MPs were sampled on quartz fiber filters. Then, all the entire filters were inserted into the pyrolysis cup, and the airborne MPs were quantified using specific pyrolysis products, as reported in the literature (Matsueda et al. 2021). The results of the quantification of numbers and mass of airborne MPs were normalized by the total volume sampled.

Results and Discussion

The workflow developed in this study proved suitable for investigating the presence of airborne MPs in indoor working environments. The integration of different methodologies provided complementary perspectives on the same samples, ensuring that both number and mass concentrations could be assessed. This multi-analytical strategy allowed not only the abundance of particles but also their polymeric composition, offering a more complete understanding of MPs occurrence in the air. Preliminary evaluations confirmed the efficiency of the digestion step as well as the suitability of Nile Red staining, as shown by recovery tests with reference materials.

Through fluorescence microscopy analysis, it was possible to obtain valuable information on the size distribution and the number of MPs collected on polycarbonate filters. Moreover, the Py–GC–MS analyses of the quartz filters contributed to quantifying the overall polymeric load in terms of mass concentration. The combined application of these methods provided a dual perspective: one focused on particle counts and classification, and the other on the quantification in mass of the plastic content in air samples. This integration is essential to obtain results that are not only descriptive but also comparable across different studies and contexts. The findings confirm that airborne MPs can be successfully identified and quantified in working areas. Variability of airborne MPs concentrations between different working areas was observed, probably indicating how different working areas can act as a source of MPs, and, in combination with limited ventilation, serve as zones of accumulation for these pollutants.

Conclusions

This study highlights the potential of an integrated analytical approach for the detection, characterization, and quantification of airborne MPs in indoor working environments. By combining fluorescence microscopy and Py–GC–MS analysis, the workflow provides information about the number and mass concentration, offering a robust framework for future monitoring studies. By Py–GC–MS analyses, it was possible to observe different concentrations of PVC, PP, PS, ABS, and SBR in different working areas. The observed variability across different working areas emphasizes the need to consider local sources and ventilation conditions in exposure assessments. Overall, these findings reinforce the importance of developing and validating standardized protocols and encourage further research to better understand the occurrence and potential health implications of MPs concentration in indoor environments.

References

Amato-Lourenço LF, dos Santos GL, Wiebeck H, Carvalho-Oliveira R, Mauad T (2022) Atmospheric microplastic fallout in outdoor and indoor environments in São Paulo megacity. Sci Total Environ 821:153450–153450

Andrady AL (2011) Microplastics in the marine environment. Mar Pollut Bull 62(8):1596–1605

ISO (2023) ISO 24187:2023—principles for the analysis of microplastics present in the environment

Jahanzaib M, Sharma S, Park D (2025) Microplastics comparison of indoor and outdoor air and ventilation rate effect in outskirts of the Seoul metropolitan city. Emerg Contam 11(1):100408

Jeong E, Lee JY, Redwan M (2024) Animal exposure to microplastics and health effects: a review. Emerg Contam 10(4):100369

Jiang W, Yan X, Lv Y (2025) A critical review on the migration, transformation, sampling, analysis and environmental effects of microplastics in the environment. J Environ Sci 154:645–664

Kalaronis D, Ainali NM, Evgenidou E, Kyzas GZ, Yang X, Bikiaris DN, Lambropoulou DA (2022) Microscopic techniques as means for the determination of microplastics and nanoplastics in the aquatic environment: a concise review. Green Anal Chem 3:100036

Käppler A, Fischer D, Oberbeckmann S, Schernewski G, Labrenz M, Eichhorn KJ, Voit B (2016) Analysis of environmental microplastics by vibrational microspectroscopy: FTIR, Raman or both? Anal Bioanal Chem 408(29):8377–8391

Matsueda M, Mattonai M, Iwai I, Watanabe A, Teramae N, Robberson W, Ohtani H, Kim YM, Watanabe C (2021) Preparation and test of a reference mixture of eleven polymers with deactivated inorganic diluent for microplastics analysis by pyrolysis-GC–MS. J Anal Appl Pyrolysis 154:104993–104993

Muñiz R, Rahman MS (2025) Microplastics in coastal and marine environments: a critical issue of plastic pollution on marine organisms, seafood contaminations, and human health implications. J Hazard Mater Adv 18:100663

O'Brien S, Rauert C, Ribeiro F, Okoffo ED, Burrows SD, O'Brien JW, Wang X, Wright SL, Thomas KV (2023) There's something in the air: a review of sources, prevalence and behaviour of microplastics in the atmosphere. Sci Total Environ 874

Sadique SA, Konarova M, Niu X, Szilagyi I, Nirmal N, Li L (2025) Impact of microplastics and nanoplastics on human health: emerging evidence and future directions. Emerg Contam 11(3):100545

Saha SC, Saha G (2024) Effect of microplastics deposition on human lung airways: a review with computational benefits and challenges. Heliyon 10(2):e24355

Sajjad M, Huang Q, Khan S, Khan MA, Liu Y, Wang J, Lian F, Wang Q, Guo G (2022) Microplastics in the soil environment: a critical review. Environ Technol Innovation 27:102408

Song J, Wang C, Li G (2024) Defining primary and secondary microplastics: a connotation analysis. ACS EST Water 4(6):2330–2332

Yakovenko N, Pérez-Serrano L, Segur T, Hagelskjaer O, Margenat H, Le Roux G, Sonke JE (2025) Human exposure to PM10 microplastics in indoor air. PLoS One 20(7):e0328011

Yin Z (2023) The pollution of microplastics in sediments: the ecological risk assessment and pollution source analysis. Sci Total Environ 859:160323

Zhao S, Kvale KF, Zhu L, Zettler ER, Egger M, Mincer TJ, Amaral-Zettler LA, Lebreton L, Niemann H, Nakajima R, Thiel M, Bos RP, Galgani L, Stubbins A (2025) The distribution of subsurface microplastics in the ocean. Nature 641(8061):51–61

Chapter 5
Bacterial Cellulose as a Sustainable Adsorbent for Pollutants and Microplastics in Water

Erika Caianiello, Federico Olivieri, Martina Aulitto, Rachele Castaldo, Roberto Avolio, Maria Emanuela Errico, Patrizia Contursi, Gennaro Gentile, and Mariacristina Cocca

Introduction

Water is essential for human life, but rapid industrial development in sectors such as textiles, paints, leather processing, pharmaceuticals, and personal care products has led to the release of large amounts of organic pollutants into aquatic environments. These contaminants include dyes, solvents, surfactants, and organometallic compounds, many of which are persistent and harmful to ecosystems and human health. Dyes are particularly problematic because they are often toxic, mutagenic, carcinogenic, and highly resistant to biodegradation. The textile industry alone releases more than 10^6 kg of dyes annually into wastewater streams, and recent advances in dye synthesis have further increased their structural stability, making them difficult to degrade after use (Iqbal et al. 2017).

To mitigate dye pollution, various wastewater treatment technologies have been explored, such as oxidation, coagulation, ion exchange, membrane separation, biological degradation, and adsorption. Among these, adsorption is considered the most promising approach due to its simplicity, high efficiency, cost-effectiveness, and minimal secondary pollution (Burevska-Atkovska et al. 2024; Dao et al. 2023; Guerritore et al. 2022; Li et al. 2020). Nevertheless, many conventional methods are ineffective against chemically stable and non-biodegradable dyes, emphasizing the need for advanced adsorbent materials.

E. Caianiello (✉) · F. Olivieri · R. Castaldo · R. Avolio · M. E. Errico · G. Gentile · M. Cocca
Institute for Polymers, Composites and Biomaterials National Research Council of Italy, Pozzuoli, (NA), Italy
e-mail: erikacaianiello@cnr.it

M. Aulitto · P. Contursi
Biology Department, University of Naples Federico II, Naples, Italy

G. Gentile et al. (eds.), *Proceedings of the 4th International Conference on Microplastic Pollution in the Mediterranean Sea*, Springer Water,
https://doi.org/10.1007/978-3-032-23847-4_5

In this context, sustainable and renewable adsorbents have gained increasing attention. In particular, cellulose is a natural, abundant and renewable polymer, valued as a sustainable alternative to synthetic polymers for its thermal stability, chemical resistance, strength, non-toxicity, and biodegradability. Various strategies have been reported in the literature to enhance its adsorption capabilities (Maglione et al. 2025). Among cellulose types, bacterial cellulose (BC), an extracellular biopolymer produced by bacteria, is particularly attractive due to its high crystallinity, three-dimensional nanofibrous structure, large surface area, and excellent mechanical properties (Dao et al. 2023). These features make BC suitable for water remediation applications, especially for dye and heavy-metal removal. However, BC typically exhibits higher affinity for cationic pollutants than for anionic dyes, mainly because the electron-rich oxygen atoms in the cellulose backbone hinder the adsorption of negatively charged species (Abdel-hamid and Mathew 2021; Dao et al. 2023).

Carbon-based materials, particularly graphene and graphene oxide (GO), are also widely studied as adsorbents due to their high surface area, porosity, and tunable surface chemistry, which enable efficient removal of both anionic and cationic dyes and other organic contaminants (Castaldo et al. 2018; Ramesha et al. 2011). However, the practical use of GO is limited by difficulties in recovery and reuse. To address this issue, GO has been incorporated into different engineered systems (Castaldo et al. 2022), including cellulose matrices, improving structural stability and adsorption performance (Ramesha et al. 2011).

Based on these findings, the present study investigates the potential of microbially synthesized bacterial cellulose for the removal of organic contaminants and microplastics from water. BC was obtained either in static cultures as a continuous pellicle (biofilms) or in stirring cultivation as BC beads. In both cases, GO was added to the bacterial cultivation to obtain hybrid samples embedding the 2D nanofillers, thereby exploiting its adsorption capacity towards various aromatic pollutants. The samples were preliminarily characterized by scanning electron microscopy (SEM) and thermogravimetric analysis (TGA). Adsorption performance towards organic pollutants was assessed though UV–Vis spectroscopy.

Materials and Methods

Bacterial cellulose was obtained either under static aerobic conditions as layers or under stirred anaerobic conditions as cellulose beads. In both approaches, GO was added during bacterial cultivation yielding hybrid materials with embedded 2D nanofillers and enhanced adsorption capability toward various aromatic pollutants. Considering both yield and uniformity of GO distribution within the cellulose matrix, the GO-containing samples that proved most promising were those obtained as beads. BC biofilm and BC-GO beads were characterized as described below.

Morphological analysis was carried out using a FEI Quanta 200 FEG SEM. Freeze-dried samples were mounted on aluminum stubs, sputter-coated, and observed with a secondary electron detector at an accelerating voltage of 10–30 kV.

Thermogravimetric analysis (TGA) was carried out by using a Perkin Elmer Pyris 1 thermogravimetric analyzer, using nitrogen as purge gas (40 mL/min). About 3 mg of sample was heated from 25 to 800 °C at 10 °C/min. In order to eliminate all the water content, the samples were held at 100 °C for 10 min before the heating continued.

BC and BC–GO were evaluated for the adsorption of Rhodamine 6G (R6G) (cationic), Methyl Orange (MO) (anionic), and 2,4-dichlorophenol (DCP, 99%) (neutral) from water through kinetic and equilibrium tests. Adsorption experiments were conducted at 25 °C using a constant sorbent-to-solution ratio of 1 mg/mL, and pollutant concentrations were monitored by UV spectroscopy. Aqueous solutions with concentrations (C_0) ranging from 5 to 250 mg/L were prepared for all pollutants. Calibration curves were obtained by UV–Vis spectroscopy by measuring the characteristic absorbance peaks at 530 nm for R6G, 510 nm for MO, and 284 nm for DCP, and pollutant concentrations were determined using the Lambert–Beer law.

Results and Discussion

SEM images of GO-BC beads highlight the structural differences between BC and BC–GO. The interconnected nanofibrous structure of cellulose is clearly visible in all samples. In the beads, the nanofiller GO sheets are uniformly distributed among the fibers.

TGA was performed to compare BC and BC–GO thermal stability and to evaluate their composition by examining the residual weight percentages. Both samples undergo thermal degradation at temperatures above 250 °C. BC–GO exhibits significantly higher residual weight compared to BC, due to the GO incorporation.

Adsorption tests were carried out using R6G, DCP, and MO as model organic pollutants with different electronic charges (Fig. 5.1).

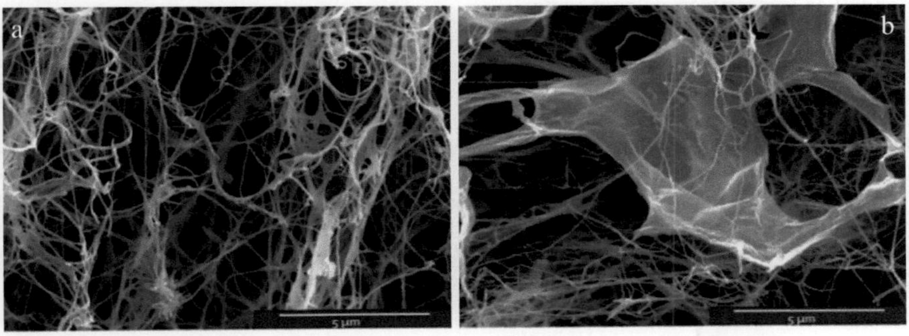

Fig. 5.1 SEM images of BC (**a**) and BC-GO (**b**)

Results demonstrated that BC–GO beads exhibited superior adsorption performance compared to BC alone. Moreover, GO incorporation allowed to adsorb pollutants that are not adsorbed with only BC. In fact, recent studies have shown that cellulose is effective in adsorbing cationic organic dyes, while it exhibits low or negligible efficiency toward anionic dyes, except for Congo Red, due to its ability to form hydrogen bonds with cellulose (Dao et al. 2023).

In this study, BC showed an R6G adsorption capacity up to 13 mg/g, while BC–GO showed significantly higher adsorption efficiency (245 mg/g) at an initial solution concentration of 250 mg/L.

BC adsorbed only a moderate amount of DCP (13 mg/g) at C_0 250 mg/L, while the hybrid sample BC–GO captured up to 130 mg/g at the same solution concentration.

Instead, BC did not adsorb any appreciable amount of the anionic dye MO at any concentration, as expected, while BC–GO adsorbed up to 22 mg/g of pollutant at C_0 250 mg/g.

Conclusions

In conclusion, these results highlight the promising potential of bacterial cellulose–based materials, especially BC–GO nanocomposites, for water purification applications. Their adsorption efficiency, versatility, and sustainable nature make them attractive candidates for advanced treatment systems.

Moreover, due to their three-dimensional fibrous structure and high porosity, BC-GO can be used to retain micro and nanoparticles, like micro and nanoplastics.

Future work will focus on further characterization and optimization of adsorption performance, evaluation of material stability and reusability, and assessment under more complex and realistic water matrices.

These studies will contribute to a deeper understanding of material behavior and support the development of effective strategies for the removal of organic pollutants as well as microplastics from water.

Acknowledgements The authors also acknowledge financial support for the projects:

• "Production of bacterial cellulose from agrifood residues by Kombucha culture: enhancing cellulose productivity and tailoring its techno-functional properties for advanced applications - BACK APP", PRIN2022 founded by European Union—Next Generation EU, C.I. 2022AEXJL7 - CUP B53D23005780006.

• "HIerarchically Porous materials with tailored ADsorption properties to face environmental pollution - HIPAD", Proposal code FIS-2023-01686, CUP B53C24009520001, financed by the Italian Science Fund (FIS 2 Call) of the Italian Ministry of University and Research (MUR), MUR Grant Assignment Decree No. n. 23314 adopted on 11/12/2024.

References

Abdelhamid HN, Mathew AP (2021) Cellulose-based materials for water remediation: adsorption, catalysis, and antifouling. Front Chem Eng 3:790314. https://doi.org/10.3389/fceng.2021.790314

Burevska-Atkovska K, Olivieri F, Avolio R, Castaldo R, Cocca M, Errico ME, Gentile G, Grozdanov A (2024) Amino-modified microporous hyper-crosslinked resins for heavy metal ions adsorption. Colloids Surfaces A: PhysChem Eng Asp 700:134720. https://doi.org/10.1016/j.colsurfa.2024.134720

Castaldo R, Lama GC, Aprea P, Gentile G, Lavorgna M, Ambrogi V, Cerruti P (2018) Effect of the oxidation degree on self-assembly, adsorption and barrier properties of nano-graphene. Microporous Mesoporous Mater 260:102–115. https://doi.org/10.1016/j.micromeso.2017.10.026

Castaldo R, Avolio R, Cocca M, Errico ME, Lavorgna M, Šalplachta J, Santillo C, Gentile G (2022) Hierarchically porous hydrogels and aerogels based on reduced graphene oxide, montmorillonite and hyper-crosslinked resins for water and air remediation. Chem Eng J 430:133162. https://doi.org/10.1016/j.cej.2021.133162

Dao NT, Bui HT, Kim Le PT, Le KA, Tuong Tran AT, Nguyen KD, Mai Nguyen HH, Ho PH (2023) Bacterial cellulose aerogels derived from pineapple peel waste for the adsorption of dyes. ACS Omega 8(37):33412–33425. https://doi.org/10.1021/acsomega.3c03130

Guerritore M, Olivieri F, Avolio R, Castaldo R, Cocca M, Errico ME, Lavorgna M, Silvestri B, Ambrogi V, Gentile G (2022) Hierarchical micro-to-macroporous silica nanoparticles obtained by their grafting with hyper-crosslinked resin. Microporous Mesoporous Mater 335:111864. https://doi.org/10.1016/j.micromeso.2022.111864

Iqbal MZ, Pal P, Shoaib M, Abdala AA (2017) Efficient removal of different basic dyes using graphene. Desalination Water Treat 68:226–235. https://doi.org/10.5004/dwt.2017.20213

Li B, Zhang Q, Pan Y, Li Y, Huang Z, Li M, Xiao H (2020) Functionalized porous magnetic cellulose/Fe_3O_4 beads prepared from ionic liquid for removal of dyes from aqueous solution. Int J Biol Macromol 163:309–316. https://doi.org/10.1016/j.ijbiomac.2020.06.280

Maglione A, Olivieri F, Avolio R, Castaldo R, Cocca M, Errico ME, Ambrogi V, Gentile G (2025) Hyper-cross-linked cellulose nanofibrils with spontaneous and reversible adsorption of aromatic pollutants from water as a valid alternative to fossil-based adsorbents. ACS Appl Mater Interfaces. https://doi.org/10.1021/acsami.5c05009

Ramesha GK, Vijaya Kumara A, Muralidhara HB, Sampath S (2011) Graphene and graphene oxide as effective adsorbents toward anionic and cationic dyes. J Colloid Interface Sci 361(1):270–277. https://doi.org/10.1016/j.jcis.2011.05.050

Chapter 6
Indoor Airborne Microplastics: Insights from a Literature Review Study

Claudia Campanale, Leonardo Barlucchi, Augusta Piccardi, Jacopo La Nasa, Francesca Modugno, Marcello De Rossi, Giordano Morelli, and Lorenzo Palumbo

Introduction

The omnipresence of plastic in modern society has made human exposure to microplastics (MPs) and nanoplastics (NPs) inevitable. These synthetic particles, defined as polymer fragments smaller than 5 mm, have been detected in virtually every environmental compartment, from remote mountain plateaus to deep ocean trenches, and within the confined spaces where humans spend most of their lives (European Commission 2023; Mitrano and Wohlleben 2020). As plastic materials fragment through mechanical wear, UV degradation, and chemical weathering, MPs are released into air, water, food, and dust, forming an inescapable component of the anthropogenic exposome (Frias and Nash 2019; Thompson et al. 2024; Prata et al. 2021).

Recent studies demonstrate that indoor environments constitute one of the most significant and persistent reservoirs of airborne MPs. Individuals inhale thousands of plastic particles per day, depending on the type of environment, ventilation, and textile use (Ageel et al. 2022; Dewika et al. 2023). Once suspended, MPs behave as respirable particulates, entering the respiratory tract and potentially reaching the alveolar region when particle diameters fall below 10 µm (Zhu et al. 2024).

Inhalation is increasingly recognized as a critical route of exposure, alongside ingestion and dermal contact (Pham et al. 2023; Cole et al. 2024). Airborne MPs,

C. Campanale (✉) · A. Piccardi · F. Modugno · L. Palumbo
Department of Environment and Health, National Institute of Health (ISS), Rome, Italy
e-mail: claudia.campanale@iss.it

L. Barlucchi · J. La Nasa
Department of Chemistry and Industrial Chemistry, University of Pisa, Pisa, Italy

M. De Rossi · G. Morelli
Laboratory of Industrial Hygiene, ASL Viterbo, Viterbo, Italy

G. Gentile et al. (eds.), *Proceedings of the 4th International Conference on Microplastic Pollution in the Mediterranean Sea*, Springer Water,
https://doi.org/10.1007/978-3-032-23847-4_6

especially nanoplastics, may deposit within bronchioles and alveoli, where clearance mechanisms could be inefficient. Empirical evidence has confirmed the presence of MPs in human lung tissue, sputum, bronchoalveolar lavage fluid, liver, colon, kidney, spleen, placenta, and even blood clots (Ragusa et al. 2021; Yang et al. 2022; Kopatz et al. 2023). Detection of MPs in maternal milk, placenta, and infant faeces could indicates transplacental transfer and early-life exposure, suggesting lifelong coexistence between humans and synthetic particles (Zuri et al. 2023).

Once internalized, MPs could interact with cells and tissues through both physical and chemical mechanisms. Their polymeric chains are metabolically stable and highly resistant to enzymatic degradation, leading to persistence within biological systems (Yang et al. 2022). Inhaled MPs could induce oxidative stress, mitochondrial dysfunction, and pro-inflammatory signalling, disrupting epithelial integrity and immune regulation (Junaid et al. 2022; Zhu et al. 2024). Furthermore, MPs' large surface area and hydrophobicity facilitate the adsorption of environmental contaminants such as heavy metals, polycyclic aromatic hydrocarbons (PAHs), and plastic additives. These co-contaminants may desorb under physiological conditions, amplifying toxicological risks (Campanale et al. 2022). MPs also serve as potential carriers for microorganisms, functioning as vectors for pathogens and endotoxins, a mechanism described as the "Trojan horse effect" (Junaid et al. 2022).

The biological response to MPs depends on multiple parameters, including polymer type, particle size, shape, surface chemistry, and concentration. Fibers, which dominate indoor air, could mechanically irritate airway tissues and stimulate chronic inflammatory processes (Zhu et al. 2024). Ultrafine MPs and NPs may translocate beyond the pulmonary barrier into systemic circulation, reaching secondary organs such as the liver, kidney, spleen, and brain (Kopatz et al. 2023). This systemic transport raises concerns about long-term impacts on metabolic, neurological, and reproductive functions.

Despite these growing concerns, research on airborne MPs in indoor environments remains fragmented and methodologically inconsistent (Kacprzak and Tijing 2022). This work synthesizes evidence from 57 observational studies published between 2017 and 2025, encompassing residential (homes, dormitories, apartments, etc.) and collective (hospitals, schools, universities, offices, public transports) indoor spaces.

It identifies major emission sources and key factors influencing airborne MP abundances in indoor spaces. The overarching objective is to promote methodological harmonization, integrate toxicological testing into future indoor air quality assessments, and ultimately improve the scientific foundation for risk management of airborne MPs.

Materials and Methods

A systematic review was performed using Web of Science, Scopus, and Google Scholar to identify 57 observational studies (2017–2025) on airborne MPs in indoor environments. Only field-based studies were included. Data were first extracted regarding sampling strategies, which were classified as active (pumped air filtration) or passive (depositional collection). Information on the analytical techniques employed was also compiled. Reported concentration results showed substantial variability, with data expressed either as MPs/m^3 for volumetric air samples or MPs/m^2/day for depositional fluxes, reflecting methodological differences across studies.

Results and Discussion

In total, 57 observational studies examining airborne MPs in indoor settings were identified across multiple continents. These studies span geographically diverse regions, including Asia, Europe, North America, South America, and Australia, reflecting a global interest in quantifying indoor MP exposure. The distribution of studies is uneven: East Asia (particularly China) constitutes a large share of published work (Fig. 6.1).

Sampling

The reviewed studies revealed substantial heterogeneity in sampling strategies for airborne MPs in indoor environments. Approximately 56% of investigations employed active sampling, which involves drawing air through filters using pumps or high-volume samplers, allowing for quantitative concentration estimates and size-selective collection. In contrast, 37% relied on passive sampling, capturing particles through natural deposition over extended exposure periods, while a smaller proportion combined both approaches. Filter materials varied widely, including PTFE, glass fiber, quartz, and cellulose nitrate membranes. Sampling durations ranged from hours (active) to several weeks (passive), and airflow rates spanned from a few to over 1000 L/min, introducing major differences in detection limits and representativeness. This methodological diversity directly affects comparability among studies and the interpretation of exposure levels.

Analytical Techniques

Analytical approaches also differed considerably, reflecting the absence of standardized protocols. The most commonly employed methods included optical microscopy for morphological identification and particle counting, μ-FTIR and μ-Raman spectroscopy for polymer characterization, and pyrolysis–gas chromatography–mass spectrometry (Py-GC-MS) for mass-based quantification. While microscopy provides rapid screening, its resolution limits detection to particles larger than ~50

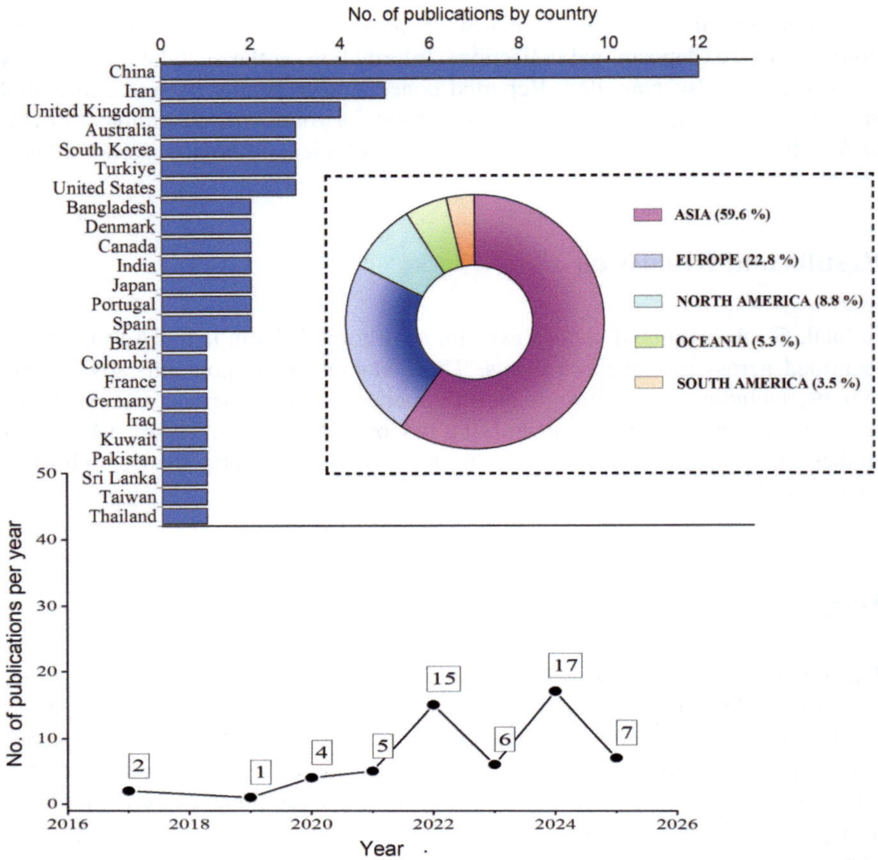

Fig. 6.1 Geographical and temporal trends of indoor airborne MP studies (2017–2025) showing publications by country, continent, and year

μm. Spectroscopic methods allow chemical confirmation of polymer types, typically identifying polyester, polyethylene, polypropylene, nylon, and polystyrene as dominant materials. However, differences in spectral resolution, pretreatment, and database matching criteria contribute to variability in reported polymer distributions. Thermal techniques, though less common, enable quantification of total polymer mass but lack particle-size resolution.

Abundance

Reported concentrations vary by several orders of magnitude due to differences in sampling techniques, analytical methods, and environmental conditions. Across all datasets, airborne MP concentrations in indoor environments were consistently higher than in outdoor controls, often by one to two orders of magnitude. Reported indoor concentrations ranged from 0.5 to 14,088 MPs/m^3 in residential spaces and 2 to 8305 MPs/m^3 in collective settings such as offices, schools, and hospitals, compared to outdoor averages near 260 MPs/m^3. Fibers represented the predominant morphology in nearly all studies, followed by fragments and films. Factors influencing MP abundance included the presence of synthetic textiles, ventilation efficiency, occupancy level, and human activity, all of which promote fiber release and resuspension. Seasonal variation was also observed, with higher concentrations often recorded during colder months when ventilation is reduced. Despite the variability, the evidence confirms that indoor air is a major exposure route to MPs.

Conclusions

Indoor environments represent a major but still underexplored source and exposure pathway for airborne MPs. Concentrations measured in residential and collective indoor spaces frequently exceed outdoor levels, confirming the strong influence of human activity, synthetic materials, and ventilation conditions. The predominance of fibrous particles, mainly polyester, polyethylene, polypropylene, and nylon, indicates that textiles and furnishings are primary emission sources.

Future research should prioritize the harmonization of sampling and analytical protocols, the adoption of standardized exposure metrics, and the inclusion of toxicological assays to clarify health implications associated with inhalation exposure. Establishing unified methods and dose–response relationships will be essential to advance risk assessment, improve air-quality monitoring, and support effective mitigation strategies aimed at reducing airborne MP exposure in built environments.

References

Ageel HK, Harrad S, Abdallah MAE (2022) Occurrence, human exposure, and risk of microplastics in the indoor environment. Environ Sci Process Impacts 24:17–31. https://doi.org/10.1039/D1EM00301A

Campanale C, Savino I, Massarelli C, Uricchio VF (2022) Assessment of metals associated with virgin pre-production and freshwater microplastics collected by an Italian river. NanoImpact 28:100438. https://doi.org/10.1016/J.IMPACT.2022.100438

Cole M, Gomiero A, Jaén-Gil A et al (2024) Microplastic and PTFE contamination of food from cookware. Sci Total Environ 929. https://doi.org/10.1016/J.SCITOTENV.2024.172577

Dewika M, Markandan K, Irfan NA et al (2023) Review of microplastics in the indoor environment: distribution, human exposure and potential health impacts. Chemosphere 324. https://doi.org/10.1016/J.CHEMOSPHERE.2023.138270

European Commission, Commission Regulation (EU) (2023/2055 of 25 September 2023) Amending Annex XVII to Regulation (EC) No 1907/2006 of the European Parliament and of the Council as Regards Synthetic Polymer Microparticles. https://Eur-Lex.Europa.Eu/Legal-Content/EN/TXT/?Uri=CELEX:32023 R2055

Frias JPGL, Nash R (2019) Microplastics: finding a consensus on the definition. Mar Pollut Bull 138:145–147. https://doi.org/10.1016/J.MARPOLBUL.2018.11.022

Junaid M, Siddiqui JA, Sadaf M et al (2022) Enrichment and dissemination of bacterial pathogens by microplastics in the aquatic environment. Sci Total Environ 830. https://doi.org/10.1016/J.SCITOTENV.2022.154720

Kacprzak S, Tijing LD (2022) Microplastics in indoor environment: sources, mitigation and fate. J Environ Chem Eng 10:107359. https://doi.org/10.1016/J.JECE.2022.107359

Kopatz V, Wen K, Kovács T et al (2023) Micro- and nanoplastics breach the blood–brain barrier (BBB): biomolecular corona's role revealed. Nano 13. https://doi.org/10.3390/NANO13081404

Mitrano DM, Wohlleben W (2020) Microplastic regulation should be more precise to incentivize both innovation and environmental safety. Nat Commun 11:1–12. https://doi.org/10.1038/s41467-020-19069-1

Pham DT, Kim J, Lee SH et al (2023) Analysis of microplastics in various foods and assessment of aggregate human exposure via food consumption in Korea. Environ Pollut 322. https://doi.org/10.1016/J.ENVPOL.2023.121153

Prata JC, da Costa JP, Lopes I, Andrady AL, Duarte AC, Rocha-Santos T (2021) A one health perspective of the impacts of microplastics on animal, human and environmental health. Sci Total Environ 777. https://doi.org/10.1016/J. SCITOTENV.2021.146094

Ragusa A, Svelato A, Santacroce C et al (2021) Plasticenta: first evidence of microplastics in human placenta. Environ Int 146. https://doi.org/10.1016/J.ENVINT.2020.106274

Thompson RC, Courtene-Jones W, Boucher J, Pahl S, Raubenheimer K, Koelmans AA (2024) Twenty years of microplastic pollution research—What have we learned? Science 386(6720). https://doi.org/10.1126/SCIENCE.ADL2746

Yang W, Jannatun N, Zeng Y et al (2022) Impacts of microplastics on immunity. Front Toxicol 4. https://doi.org/10.3389/FTOX.2022.956885

Zhu L, Kang Y, Ma M et al (2024) Tissue accumulation of microplastics and potential health risks in human. Sci Total Environ 915. https://doi.org/10.1016/J.SCITOTENV.2024.170004

Zuri G, Karanasiou A, Lacorte S (2023) Human biomonitoring of microplastics and health implications: a review. Environ Res 237. https://doi.org/10.1016/J.ENVRES.2023.116966

Chapter 7
Turning 'Ghost Boats' into Blue Opportunities: Chemical Fingerprinting and Sustainable Solutions for End-of-Life Vessels in Venice

Teresa Cecchi and Davide Poletto

Introduction

Venice, a UNESCO World Heritage Site, is among the most iconic and densely visited urban landscapes in the world. Yet, this unique lagoon ecosystem is increasingly exposed to human-induced pressures. The intense movement of visitors, together with heavy freight and leisure boat traffic, introduces multiple and cumulative pollutants into the lagoon's semi-enclosed waters (Cecchi 2021). Among the emerging forms of marine litter in the area are the so-called *"ghost boats"*—abandoned end-of-life (EOL) vessels whose owners are unknown. Abandoned recreational boats, or "ghost boats," have become a significant yet overlooked source of pollution in the Venice Lagoon and worldwide. As these fibreglass vessels deteriorate, they release particles and volatile compounds that can affect both ecosystems and human health. This study provides an integrated chemical assessment of pollutants originating from decaying hulls, combined with an evaluation of a circular solution for their end-of-life recovery. A green headspace solid phase microextraction gas chromatography mass spectrometry (HS-SPME-GC-MS) method was used to analyse water, mud, and soil samples collected near an abandoned-boat hotspot. To explore safer disposal pathways, the work also examined panels produced via a patented mechanical recycling process that transforms dismantled boats into new composite materials. Independent

T. Cecchi (✉)
ITT Montani, Fermo, Fermo, Italy
e-mail: cecchi.teresa@istitutomontani.edu.it

D. Poletto (✉)
Venice Lagoon Plastic Free, Venice, Italy
e-mail: d.poletto@plasticfreevenice.org

G. Gentile et al. (eds.), *Proceedings of the 4th International Conference on Microplastic Pollution in the Mediterranean Sea*, Springer Water,
https://doi.org/10.1007/978-3-032-23847-4_7

testing by CATAS detected no VOC emissions above instrumental limits, demonstrating that the recycled fiberglass product is chemically inert and suitable for low-emission applications. Overall, the study highlights the dual relevance of ghost boats as a source of emerging contaminants and as an opportunity for circular-economy innovation. The combined analytical and recycling evidence supports the development of local and EU-level strategies for sustainable governance models of end-of-life vessels, turning a persistent environmental common bad into a new economic pattern within the frame of the sustainable blue circular economy paradigm.

Materials and Methods

Sampling Locations

Three distinct microenvironments, identified during a routine Venice Lagoon Plastic Free (VLPF) monitoring patrol, were sampled in triplicate within the Canale Saccagnana area, located in the municipality of Cavallino Treporti (Fig. 7.1). The selected points correspond to:

(A) water collected inside an abandoned fiberglass boat, where floating algae were visible;
(B) mud directly touching the submerged portion of the vessel;
(C) soil in contact with the boat's hull.

On the day of the cleanup intervention, VLPF staff collected three replicate samples from each microenvironment (A, B, and C). Samples were placed in labelled glass containers at the location indicated in Fig. 7.1 (GPS coordinates 45.4646733, 12.4697035) and put in labelled glass containers, fitted with screw caps. All sampling activities were conducted under dry-weather conditions, defined as at least 48 hours without precipitation. Specimens were transported on ice to the laboratory and stored at − 18 °C until analytical processing.

GC-MS Analysis

To identify volatile organic compounds (VOCs), 5.000 mL of water or 5.000 ± 0.010 g of mud or soil were put in a 20 mL headspace vial, closed by a PTFE/silicone septum. Volatile compound standards, when commercially available, were purchased from Sigma-Aldrich (Milan, Italy). Volatiles in the headspace were measured in triplicate using HS-SPME-GC-MS (Cecchi 2025).

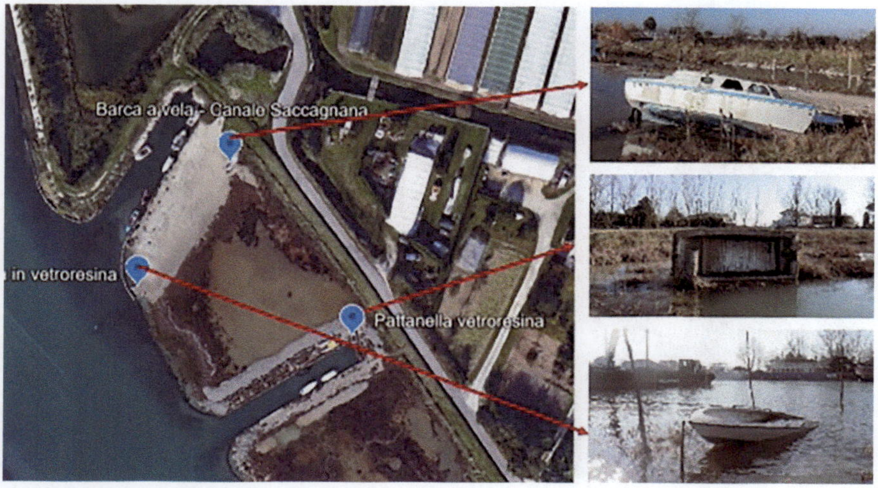

Fig. 7.1 Sampling microenvironments

Results and Discussion

Recreational boats in Italy and across Europe have historically been designed without considering end-of-life treatment. Consequently, many deregistered or unregistered vessels have not been dismantled properly and often end up abandoned in natural areas (ICOMIA 2020).

A core issue is the absence of a coherent European framework for small-vessel disposal. Neither the Waste Framework Directive, the Recreational Craft Directive, nor the Ship Recycling Regulation fully addresses it. This regulatory gap results in inconsistent national practices and limited traceability, with costs frequently falling on local authorities. Proposed improvements include extending ship-recycling obligations to smaller craft, integrating design-for-disassembly into boat manufacturing, creating a dedicated waste category for boats, and introducing harmonised Extended Producer Responsibility schemes (EBI 2023).

In the Venice Lagoon, the scale of abandonment is unclear, but with about 37,000 registered recreational vessels over 10 HP, even a small yearly share reaching end-of-life could result in several hundred ghost boats. These vessels accumulate in remote canals and wetlands, worsening environmental degradation and creating long-lasting pollution hotspots.

To counter this trend, VLPF launched the "Ghost Boats" Program (now entering its third operational cycle) with initial support from the EU H2020 project MAELSTROM, which is now expanded in the Adriatic-Ionian Region thanks to the EU Interreg IPA ADRION JOINABLE project. In partnership with GEES Recycling (Aviano, Italy), the initiative established a circular chain that repurposes abandoned fiberglass boats into new composite materials (Colledani and Turri 2022). The

Fig. 7.2 Summing the EOL boats circular pilot pattern from removal of illegally dumped 'Ghost Boats' to shredding, and end products (e.g. a bench) CATAS test

patented recycling process (EP2844445) includes dismantling hulls, separating materials, and producing a uniform fiberglass granulate. With a minimal binder content (<5%), the material is hot-pressed into durable panels suitable for green-procurement applications (Fig. 7.2). Safety tests on recycled panels performed by CATAS found no VOC emissions above analytical detection limits, confirming that the recycled composite is chemically inert and suitable for low-emission uses.

Chemical analyses conducted in this study show that deteriorating fiberglass vessels emit a complex mixture of pollutants. Forty-three VOCs were detected in sediments, water, and soil near a ghost-boat site, including polyurethane-related markers, hydrocarbons, and both anthropogenic and biogenic compounds (Fig. 7.3).

These findings confirm that abandoned boats act as continuous sources of particulate (Ciocan et al. 2020) and chemical contamination. We detected several ghost-boat-related analytes that are highly reactive in water—such as toluene diisocyanate—indicating ongoing release from polyurethane materials. Their persistence suggests that full environmental recovery cannot occur unless these pollution sources are removed. The analytical procedure applied in this study demonstrated excellent performance in terms of greenness, achieving high scores across two independent metrics (AGREE Prep: 0.8/1; Analytical Eco-Scale: 91/100), further supporting the adoption of green methodologies for environmental pollution assessment.

Overall, this work highlights two key points: abandoned boats are a notable and yet underestimated source of environmental contamination, and circular-economy solutions offer a viable pathway to transform this waste stream into safe, high-value products. Strengthening EU legislation, notably through harmonised collection systems and EPR mechanisms, would facilitate broader adoption of such practices across coastal and island regions.

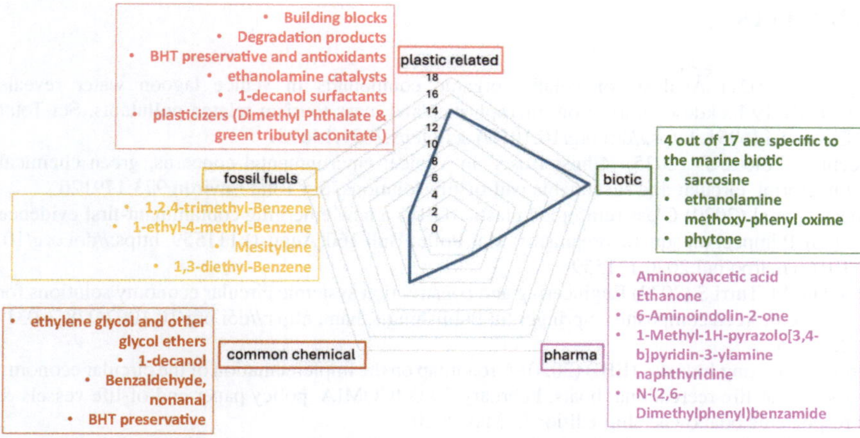

Fig. 7.3 Radar chart classification of identified analytes

Conclusions

This study provides a consolidated assessment of abandoned recreational boats as an overlooked source of pollution in the UNESCO property of Venice and its Lagoon. Using green analytical methods and a participatory field approach, we identified 43 volatile organic compounds associated with the degradation of fibreglass vessels, confirming their role as both particulate and contaminant emitters at the molecular level. The evidence highlights that ghost boats pose environmental and health risks that go far beyond visual degradation or navigational hazards. Their presence underscores the urgent need for systematic removal and sustainable end-of-life management.

The VLPF Ghost Boats Program, developed with GEES Recycling and originating from independent and EU-funded initiatives, demonstrates how civil-society action can complement or fill the policy-regulatory-monitoring gaps left by local-central administrations, and build regimes (Young 2003) that address structured common bads arising from wild dumping practices linked to pollutant dispersal in the environment (Vogler 2003). By converting abandoned vessels into inert, certified, and carbon-negative composite materials, the initiative offers a practical example of a blue circular economy within a UNESCO property (UNESCO 1987). To scale such solutions, harmonised EU legislation is essential. Extending ship-recycling rules to small craft and introducing precise Extended Producer Responsibility mechanisms would support consistent and sustainable management of end-of-life boats across Member States. Transforming "ghost boats" into "blue opportunities" ultimately depends on aligning science and technology with institutions and societal engagement to prevent and remediate this emerging contaminant pollution while developing innovative, resilient practices for hard-to-recycle plastics.

References

Cecchi T (2021) Analysis of volatile organic compounds in venice lagoon water reveals COVID-19 lockdown impact on microplastics and mass tourism related pollutants. Sci Total Environ:146951. https://doi.org/10.1016/j.scitotenv.2021.146951

Cecchi T, Poletto D (2025) 'Ghost Boats' in Venice: environmental concerns, green chemical fingerprint, circular and sustainable end-of-life solutions. Sci Total Environ 973:179126

Ciocan C et al (2020) Glass reinforced plastic (GRP) a new emerging contaminant-first evidence of GRP impact on aquatic organisms. Mar Pollut Bull 160(August):111559. https://doi.org/10.1016/j.marpolbul.2020.111559

Colledani M, Turri S (2022) Engineering and construction systemic circular economy solutions for fiber reinforced composites. Springer Int Publishing, Cham. https://doi.org/10.1007/978-3-031-22352-5

European Boating Industry (EBI) (2020) A roadmap on the implementation of the circular economy for end-of-life recreational boats. February 2023 ICOMIA, policy paper end-of-life vessels & responsible boat ownership-edition 2: May 2020

European Boating Industry (EBI) (2023) A roadmap on the implementation of the circular economy for end-of-life recreational boats

ICOMIA (2020) Policy paper end-of-life vessels & responsible boat ownership-edition 2: May 2020

UNESCO (1987) Decision 13 COM IX.22, UNESCO 1987

Vogler J (2003) Taking institutions seriously: how regime analysis can be relevant to multilevel environmental governance. Glob Environ Polit 3(2):25–39

Young OR (2003) Environmental governance: the role of institutions in causing and confronting environmental problems. Int. Environ Agreements: Polit, Law Econ 3:377–393

Chapter 8
Efficient Removal of Dyes Using Bio-derived Microporous Polymers

Antonia Cerbone, Federico Olivieri, Roberto Avolio, Mariacristina Cocca, Maria Emanuela Errico, Pierluigi Lasala, Khalid Talha, Marica Tancredi, Maria Lucia Curri, Elisabetta Fanizza, Rachele Castaldo, and Gennaro Gentile

Introduction

Synthetic dyes are widely employed in several industrial sectors, including textiles, leather processing, paper manufacturing, cosmetics, food packaging, and plastics. It is estimated that over 10,000 different dyes are commercially available, with annual production exceeding 7×10^5 tons worldwide (Sahu and Poler 2024). A significant fraction of these dyes is released into wastewater streams during industrial processing, leading to severe environmental and health concerns.

Dyes are often characterized by high chemical stability, complex aromatic structures, and resistance to biodegradation. Their persistence in aquatic environments can hinder light penetration, disrupt photosynthetic activity, and generate toxic, mutagenic, or carcinogenic by-products (Kusumlata et al. 2024). Conventional wastewater

A. Cerbone · F. Olivieri · R. Avolio · M. Cocca · M. E. Errico · R. Castaldo · G. Gentile
Institute of Polymers, Composites and Biomaterials (IPCB) National Research Council of Italy, Pozzuoli, Italy

A. Cerbone (✉)
Department of Chemical Science, University of Naples, Naples, Italy
e-mail: antoniacerbone@cnr.it

P. Lasala · K. Talha · M. Tancredi · M. L. Curri · E. Fanizza
Department of Chemistry, University of Bari A. Moro, Bari, Italy

P. Lasala · M. L. Curri · E. Fanizza
Institute for chemical and physical processes (IPCF) National Research Council of Italy, Bari, Italy

M. L. Curri · E. Fanizza
National Interuniversity Consortium of Materials Science and Technology, INSTM, Bari, Italy

G. Gentile et al. (eds.), *Proceedings of the 4th International Conference on Microplastic Pollution in the Mediterranean Sea*, Springer Water,
https://doi.org/10.1007/978-3-032-23847-4_8

treatment methods, such as biological degradation or coagulation, are frequently ineffective for dye removal, especially for azo and cationic dyes (Parida et al. 2025). In this context, adsorption-based technologies have emerged as one of the most efficient and versatile approaches for dye removal due to their simplicity, cost-effectiveness, and broad applicability (Kassa et al. 2025).

Among adsorbent materials, porous polymers have attracted increasing attention owing to their high surface area, tunable pore structure, and chemical versatility. In recent years, bio-derived microporous polymers have gained particular interest as sustainable alternatives to fossil-based materials, combining high adsorption efficiency with reduced environmental impact (Kuppusamy et al. 2026).

Results and Discussion

This contribution reviews the state of the art of bio-derived microporous polymeric systems for dye removal, with particular focus on material classification, synthesis strategies, and characterization techniques relevant to adsorption performance.

Classes of Bio-derived Microporous Polymers

Natural polysaccharides, including cellulose, chitosan, starch, and alginates, are widely explored precursors for the fabrication of microporous systems. Chitosan is among the most extensively studied materials due to its intrinsic amino groups, which promote strong interactions with dyes via electrostatic attraction and hydrogen bonding (Granados-Carrera et al. 2025, Huie et al. 2023, Kuang et al. 2023).

Lignin and lignocellulosic waste streams represent key renewable precursors for the production of microporous carbons and hyper-crosslinked polymers. Carbonization followed by physical or chemical activation yields carbon materials with extremely high specific surface areas and well-developed microporosity (Li et al. 2025). Alternatively, these precursors can be converted into hyper-crosslinked polymer networks via Friedel–Crafts reactions, generating rigid, microporous frameworks. These materials are typically employed for the adsorption of organic contaminants in water and air, including dyes (Liu et al. 2021, Yin et al. 2023, Maglione et al. 2025).

Proteins and animal-derived polymers, such as silk fibroin, gelatin, and keratin, can also be processed into microporous membranes or hierarchically porous structures. Their intrinsic biocompatibility, combined with high surface area and functional diversity, makes them attractive also for biomedical applications as well as gas storage and pollutant adsorption (Li et al. 2022, Chen et al. 2020, Gao et al. 2018). Although less explored than polysaccharides and lignin, protein-based microporous systems show promising dye adsorption performance.

Synthesis Strategies for Microporosity Development

The development of microporosity in bio-derived polymers can be achieved through a variety of chemical, physical, and thermo-chemical approaches, each characterized by distinct pore formation mechanisms and scalability potential.

Crosslinking is one of the most effective methods to induce and stabilize highly porous polymeric networks. The crosslinking density and the nature of the crosslinking agent strongly influence pore size distribution, mechanical integrity, and chemical resistance. Properly designed crosslinked systems can achieve stable microporosity suitable for repeated adsorption–desorption cycles (Maglione et al. 2025, Gunathilake et al. 2017).

Microporosity can also be achieved through sol–gel processes followed by solvent removal via freeze-drying or supercritical drying. Gelation leads to the formation of a three-dimensional, highly interconnected polymer network, while the drying step removes the solvent without collapsing the pore structure. Supercritical drying is particularly effective in preserving both micro- and mesoporosity, whereas freeze-drying often results in more irregular but still highly porous architectures (Kenar 2017).

Carbonization of biomass or bio-derived polymers under inert atmosphere at elevated temperatures converts organic precursors into carbonaceous structures. Subsequent chemical or physical activation creates additional micropores through controlled corrosion of the carbon matrix, resulting in materials with exceptionally high surface areas and adsorption capacities (Correa and Kruse 2018).

Characterization Techniques

The characterization of bio-derived microporous systems primarily focuses on evaluating specific surface area, pore size distribution, and surface chemistry. Volumetric adsorption analyzers are commonly employed to assess surface area and adsorption capacity at low relative pressures. Porosimetry techniques allow quantification of micro-, meso-, and macroporous fractions. Spectroscopic methods such as FTIR and solid-state NMR provide insights into chemical functionalities, while scanning and transmission electron microscopy enable direct observation of porous morphology and structural homogeneity.

Conclusions

Bio-derived microporous polymers represent a promising and sustainable class of materials for the efficient removal of dyes from aqueous environments. Their renewable origin, tunable porosity, and chemical versatility enable high adsorption performance while addressing environmental concerns associated with fossil-based adsorbents. The diversity of available precursors, combined with multiple synthesis and activation strategies, allows fine control over pore structure and surface functionality. Continued research efforts focused on scalability, regeneration, and real wastewater testing are expected to further enhance the practical applicability of these materials in environmental remediation.

Acknowledgments The authors acknowledge financial support under the National Recovery and Resilience Plan (NRRP), Mission 4, Component 2, Investment 1.1, Call for tender No. 1409 published on 14.09.2022 by the Italian Ministry of University and Research (MUR), funded by the European Union—NextGenerationEU– Project Title "Photocatalytically-regenerable hierarchically porous adsorbents for efficient water treatment" (PHOTOPAD, code P2022FP2W4, 2023-2025), CUP B53D23027540001, Grant Assignment Decree No. 1389 adopted on 01.09.2023 by the Italian MUR.

The authors also acknowledge financial support under the Italian Science Fund (FIS 2 Call) funded by the Italian Ministry of University and Research (MUR) for the project "HIerarchically Porous materials with tailored ADsorption properties to face environmental pollution" (HIPAD, Proposal code FIS-2023-01686, CUP B53C24009520001, Grant Assignment Decree No. n. 23314 adopted on 11/12/2024 by the Italian MUR).

References

Chen W, Li F, Chen L, Zhang Y, Zhang T, Wang T (2020) Fast self-assembly of microporous silk fibroin membranes on liquid surface. Int J Biol Macromol 156:633–639

Correa CR, Kruse A (2018) Biobased functional carbon materials: production, characterization, and applications—a review. Materials 11(9):1568

Gao A, Guo N, Yan M, Li M, Wang F, Yang R (2018) Hierarchical porous carbon activated by $CaCO_3$ from pigskin collagen for CO_2 and H_2 adsorption. Micropor Mesopor Mat 260:172–179

Granados-Carrera CM, Castro-Criado D, Abdullah JAA, Jimenez-Rosado M, Perez-Puyana VM (2025) Aerogels based on chitosan and collagen modified with fe_2o_3 and fe_3o_4 nanoparticles: fabrication and characterization. Polymers 17(2):133

Gunathilake TMSU, Ching YC, Ching KY, Chuah CH, Abdullah LC (2017) Biomedical and microbiological applications of bio-based porous materials: a review. Polymers 9(5):160

Huie J, Suqiu Z, Haiyan J, Zhijian L, Lijuan C, Nihao L, Xinhua L (2023) Microporous chitosan/polyvinyl alcohol based active packaging materials with integrated gas-transmission, radiation-cooling, anti-microbial, and ultraviolet shielding features. Chem Eng J 473:145432

Kassa A, Engida A, Endaye M (2025) Eco-friendly adsorbents for industrial dye removal: a comprehensive review of low-cost alternatives. DWT 323:101362

Kenar JA (2017) Porous structures from bio-based polymers via supercritical drying. In: Yang Y, Yu J, Xu H, Sun B (eds) Porous lightweight composites reinforced with fibrous structures. Springer, Berlin, Heidelberg

Kuang J, Cai T, Dai J, Yao L, Liu F, Liu Y, Shu J, Fan J, Peng H (2023) High strength chitin/chitosan-based aerogel with 3D hierarchically macro-meso-microporous structure for high-efficiency adsorption of Cu(II) ions and Congo red. Int J Biol Macromol 230:123238

Kuppusamy B, Mohamed Ismail FR, Balakrishnan P, Kim SC, Asrafali SP, Periyasamy T (2026) From biomass to adsorbent: a comprehensive review on bio-derived carbons for dye removal. Polymers 18(2):180

Kusumlata AB, Kumar A, Gautam S (2024) Sustainable solutions: reviewing the future of textile dye contaminant removal with emerging biological treatments. Limnol Rev 24(2):126–149

Li H, He J, Yao J (2025) A review of recent advances in biomass-derived porous carbon materials for CO_2 capture. C 11(4):92

Li J, Luo X, Lv Z, Qiang H, Hou C, Liu K, Meng C, Zhang Y, Liu F, Zhang B (2022) Microporous structures on mineralized collagen mediate osteogenesis by modulating the osteo-immune response of macrophages. Front Bioeng Biotechnol 10

Liu N, Chen J, Wu Z, Zhan P, Zhang L, Wei Q, Wang F, Shao L (2021) Construction of microporous lignin-based hypercross-linked polymers with high surface areas for enhanced iodine capture. ACS Appl Polym Mater 3(4):2178–2188

Maglione A, Olivieri F, Avolio R, Castaldo R, Cocca M, Errico ME, Ambrogi V, Gentile G (2025) Hyper-cross-linked cellulose nanofibrils with spontaneous and reversible adsorption of aromatic pollutants from water as a valid alternative to fossil-based adsorbents. ACS Appl Mater Interfaces 17(26):38619–38634

Parida VK, Singh N, Priyadarshini M, Kumari P, Datta D, Tambi A (2025) Insights into the synthetic dye contamination in textile wastewater: impacts on aquatic ecosystems and human health, and eco-friendly remediation strategies for environmental sustainability. J Ind Eng Chem 150:247–264

Sahu A, Poler JC (2024) Removal and degradation of dyes from textile industry wastewater: benchmarking recent advancements, toxicity assessment and cost analysis of treatment processes. J Environ Chem Eng 12(5):113754

Yin L, Hu P, Liang C, Wang J, Li M, Qu W (2023) Construction of self-supporting ultra-micropores lignin-based carbon nanofibers with high areal desalination capacity. Int J Biol Macromol 225:1415–1425

Chapter 9
Bacterial Cellulose–Poly(Vinyl Alcohol) Foams for Microplastic Retention in Water Filtration

Raffaele Condò, Erika Caianiello, Roberto Avolio, Rachele Castaldo, Maria Emanuela Errico, Gennaro Gentile, Federico Olivieri, and Mariacristina Cocca

Introduction

Microplastics (MPs) are defined as plastic fragments under 5 mm, with diverse shapes, chemical compositions, and densities. Laundering synthetic clothes is a major MPs source for aquatic environments, as they are released from textiles, pass through wastewater treatment, and contaminate marine ecosystems (Santonicola et al. 2025). Ingesting MPs poses risks to aquatic organisms and potentially to human health (Volgare et al. 2022). Numerous technologies for MPs removal exist, including physical, chemical, and biological methods, though some are not entirely "green" (Yang et al. 2023).

This work shows porous foams based on bacterial cellulose (BC), obtained from Kombucha fermentation, and poly(vinyl alcohol) (PVA), a biodegradable polymer, for the adsorption and removal of MPs from laundry wastewater.

Materials and Methods

Foams were produced using a sustainable technique that forms a porous network by entrapping air inside a water-based polymer/filler dispersion through high-speed stirring. Various formulations were obtained by adjusting the relative amounts of BC

R. Condò · E. Caianiello · R. Avolio · R. Castaldo · M. E. Errico · G. Gentile · F. Olivieri · M. Cocca (✉)
Institute of Polymers, Composites and Biomaterials, National Research Council of Italy, Pozzuoli, Italy
e-mail: mariacristina.cocca@cnr.it

G. Gentile et al. (eds.), *Proceedings of the 4th International Conference on Microplastic Pollution in the Mediterranean Sea*, Springer Water,
https://doi.org/10.1007/978-3-032-23847-4_9

Table 9.1 Foams composition

1	99 wt% PVA and 1 wt% cellulose	PVA_CA/BC 1
2	95 wt% PVA and 5 wt% cellulose	PVA_CA/BC 5
3	5 wt% PVA and 95 wt% cellulose	PVA_CA/BC 95
4	1 wt% PVA and 99 wt% cellulose	PVA_CA/BC 99

and PVA, crosslinked with citric acid (CA). In particular, four different compositions were developed as listed in Table 9.1. For all the foams, CA was added at 30 phr.

Foams internal structure was observed by scanning electron microscopy (SEM), Filtration experiments were conducted to determine the efficiency of the foams in removing microplastics from washing effluents containing a known concentration of MPs released by polyester textiles using a GyroWash apparatus (De Falco et al. 2018).

Results and Discussion

The morphologies of the obtained foams were strongly affected by the amount of BC. Foams with low BC content have a regular structure with spherical interconnected large and small pores. In contrast, foams with high BC amount have a completely different structure, characterized by a disordered and interconnected network of fibres.

During filtration tests, foams with higher PVA content released measurable amounts of polymer into the filtrate, which interfered with microfiber counting and indicated the release of MPs from the foam itself. This behavior represents an intrinsic limitation for the use of high-PVA-content foams in MP retention applications. In contrast, foams with higher cellulose content exhibited high microfiber trapping efficiency, reaching a maximum removal efficiency of 75%. BC provides an excellent matrix for MP entrapment, while the incorporation of a PVA phase improves the mechanical integrity of the material, enabling its practical application in foam form (Fig. 9.1).

Conclusions

Sustainable foams were obtained using a biodegradable polymer, PVA, and natural fibres, BC, to be used to capture microplastics released from textile to water. Future work should focus on improving the stability of these foams in water environments and maximizing the interaction between PVA and BC phases to guarantee better integrity and handing.

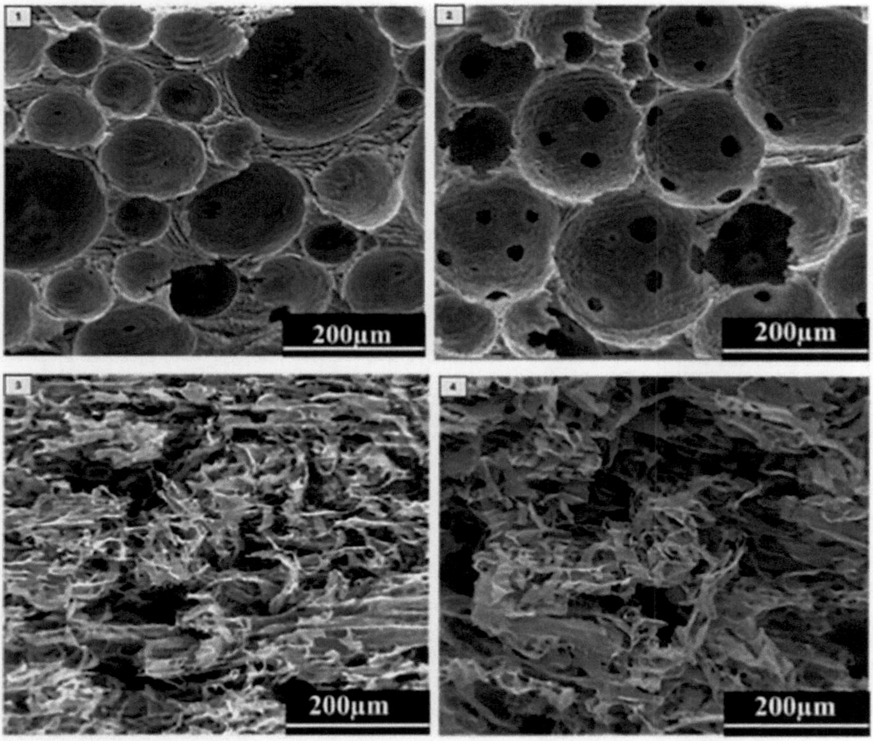

Fig. 9.1 SEM micrographs of: **1**) PVA_CA/BC 1; **2**) PVA_CA/BC 5; **3**) PVA_CA/BC 95; **4**) PVA_CA/BC 99. PVA_CA/BC and PVA_CA/BC 5 display spherical pores of varying size. PVA_CA/BC 95 and PVA_CA/BC 99 depict a fibrous, layered texture

References

De Falco F, Gullo MP, Gentile G, Di Pace E, Cocca M, Gelabert L, Brouta-Agnésa M, Rovira A, Escudero R, Villalba R, Mossotti R, Montarsolo A, Gavignano S, Tonin CM (2018) Environ Pollut 236:916–925

Santonicola S, Volgare M, Olivieri F, Cocca M, Colavita G (2025) Natural and regenerated cellulosic microfibers dominate anthropogenic particles ingested by commercial fish species from the Adriatic Sea. Foods 14(7):1237. https://doi.org/10.3390/foods14071237

Volgare M, Santonicola S, Cocca M, Avolio R, Castaldo R, Errico ME, Gentile G, Raimo G, Gasperi M, Colavita G (2022) Sci Rep 12(1):21827

Yang J, Monnot M, Sun Y, Asia L, Wong-Wah-Chung P, Doumenq P, Moulin P (2023) Water Res 232:119673

Chapter 10
Detection of Microplastics in Environmental Compartments Near a Traffic-Intensive Road

Sara Criollo, Lucian Iordachescu, Subhash S. Rathnaweera, Jeanette Lykkemark, Lelum Manamperuma, Eilen A. Vik, Luis Camacho, and Jes Vollertsen

Introduction

Microplastic (MP) pollution has become a global concern due to its occurrence in diverse ecosystems, including marine (Muñiz and Rahman 2025; Zúñiga Umaña et al. 2025), freshwater (Shen et al. 2021; Sunny et al. 2025), terrestrial (Sahai et al. 2025), and even atmospheric environments (Sridharan et al. 2021). MPs, defined as plastic fragments < 5 mm, include primary plastics, intentionally manufactured for products, and secondary plastics derived from fragmentation of larger items (Barnes et al. 2009). While marine MPs have been widely studied, knowledge on their occurrence in road dust, road runoff, and air remains limited (Horton et al. 2017; Morioka et al. 2023).

Road traffic is a key source of MPs. Road dust, generated by vehicle abrasion and surrounding deposition, contributes significantly to MP loads in stormwater runoff (Liu et al. 2019; Monira et al. 2022). Tyre wear particles (TWPs) constitute a major fraction, alongside brake wear and road marking plastics (Giechaskiel et al. 2024). Once emitted, MPs disperse via air, water, and solid surfaces, creating risks for ecosystems and human health. Stormwater sediments often show

S. Criollo (✉) · L. Iordachescu · J. Lykkemark · L. Camacho · J. Vollertsen
Section of Civil and Environmental Engineering, Department of the Built Environment, Aalborg University, Aalborg Øst, Denmark
e-mail: saraccdlc@build.aau.dk

S. Criollo
Civil and Environmental Engineering Department, Universidad de los Andes, Bogotá D.C., Colombia

S. S. Rathnaweera · L. Manamperuma · E. A. Vik
Aquateam COWI AS, Oslo, Norway

G. Gentile et al. (eds.), *Proceedings of the 4th International Conference on Microplastic Pollution in the Mediterranean Sea*, Springer Water,
https://doi.org/10.1007/978-3-032-23847-4_10

the highest MP concentrations near roads, with TWPs as dominant contributors (Rasmussen et al. 2024). Quantification approaches include micro-Fourier Transform Infrared Spectroscopy (μFTIR), effective for polymer identification and morphology (Vianello et al. 2013), though limited for rubber. Pyrolysis–GC/MS complements by quantifying TWPs, but only in terms of mass (Kirstein et al. 2021).

This study investigates MPs and TWPs in road dust, runoff, and air along a heavily trafficked Norwegian highway, considering meteorological influences on airborne MPs. A multi-matrix perspective provides insight into emission pathways and supports improved understanding of traffic-derived microplastic pollution.

Materials and Methods

This study was conducted along the E18 highway in Bamble, Norway (59.032822, 9.677411), a site equipped with a drainage station for runoff collection and continuous meteorological monitoring. Road runoff was sampled from a mixing tank via the Universal Filtering Unit (10 μm stainless-steel filters), yielding three samples between October 2022 and April 2023. Air was collected with four passive Particle Vision Sigma-2 samplers installed 17 m from the highway between March and June 2023. Road dust was obtained near the Bamble tunnel using a stainless-steel vacuum device, "Dusty" (Iordachescu et al. 2024), across four 1 m^2 plots. Samples underwent matrix-specific preparation, including density separation with sodium polytungstate, oxidation (Fenton's reagent), and ethanol concentration. Road dust was sieved (< 500 μm fraction) before processing.

For polymer characterization, subsamples were analysed with μFTIR (Cary 620 microscope, 5.5 μm resolution), providing particle counts, sizes, and estimated masses. Each air, runoff, and dust sample was scanned in triplicate aliquots. As μFTIR cannot reliably detect tyre rubber, TWPs were quantified using Py-GC/MS (Frontier Labs pyrolyzer coupled to Thermo Fisher ISQ MS), targeting 4-vinylcyclohexene as marker compound. Results from μFTIR are reported as MPs, while Py-GC/MS data are reported as TWP mass.

Results and Discussion

Road dust exhibited the highest concentrations of MPs and TWPs, confirming its role as a primary reservoir of traffic-derived plastics. MP counts ranged from 567 to 4250 particles/m^2 (31–291 μg/m^2), with TWPs reaching up to 178,777 μg/m^2. Road runoff presented intermediate levels (65–598 counts/L; 0.4–11.7 μg/L MPs; 281–1470 μg/L TWPs), while air showed the lowest burdens (5–26 counts/day; < 1 μg/day), below the TWP detection limit. Polymer distributions reflected this contrast: polypropylene dominated in road dust and runoff, whereas polyester and polyamide were more frequent in air samples.

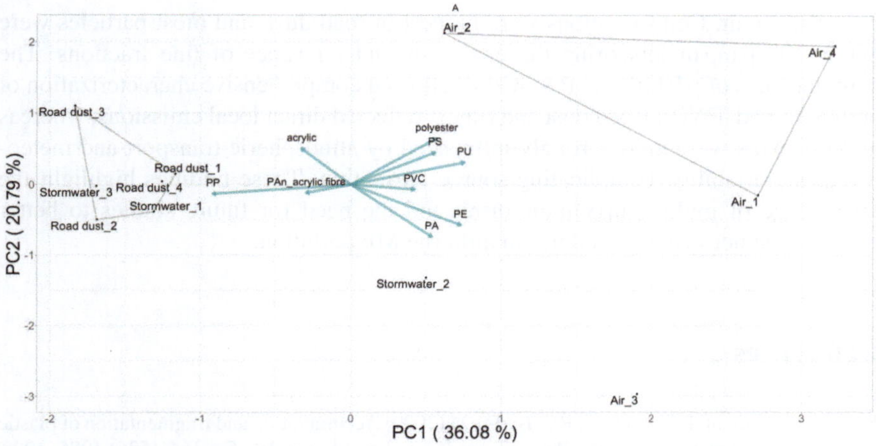

Fig. 10.1 Principal Component Analysis (PCA). *Note* This figure has been used in the submitted paper related to the research described in this document (under review and publish as a preprint) (Criollo et al. 2026)

Meteorological conditions influenced airborne MPs. Higher wind speeds coincided with lower MP capture, suggesting enhanced dispersion, while precipitation showed no consistent relationship. This highlights the difficulty of linking airborne MPs directly to local emissions, as atmospheric transport and weather variability modulate deposition patterns. In contrast, dust and runoff more directly reflected local traffic-related inputs.

Particle size analysis across all compartments revealed a predominance of small MPs (< 100 μm), particularly in road dust, which peaked between 40 and 80 μm. Such fine particles are of concern given their greater mobility and potential for environmental and biological interactions.

The Principal Component Analysis (Fig. 10.1) integrated these patterns: air samples separated clearly from dust and runoff, reflecting distinct polymer fingerprints and atmospheric influences. Dust and runoff clustered more closely, dominated by polypropylene and indicative of common traffic-related sources. This separation underscores the value of a multi-matrix approach for distinguishing emission pathways and assessing the relative roles of local deposition versus atmospheric transport.

Conclusion

This study demonstrated that traffic-related microplastics (MPs) and tyre wear particles (TWPs) are present in road dust, runoff, and air along a Norwegian highway, with polypropylene dominating in dust and runoff, while polyester and polyamide prevailed in air. Concentrations were highest in road dust, and most particles were below 100 μm, underscoring the environmental relevance of fine fractions. The combined use of μFTIR and Py-GC/MS allowed comprehensive characterization of both MPs and TWPs. Road dust and runoff reflected direct local emissions, whereas airborne MPs were more strongly influenced by atmospheric transport and meteorological variability, complicating source attribution. These findings highlight the importance of multi-matrix monitoring and the need for future studies to better capture weather-driven dynamics in airborne MP pollution.

References

Barnes DK, Galgani F, Thompson RC, Barlaz M (2009) Accumulation and fragmentation of plastic debris in global environments. Philos Trans R Soc Lond Ser B Biol Sci 364(1526):1985–1998

Criollo S, Iordachescu L, Rathnaweera SS, Lykkermark J, Manamperuma L, Vik EA, Camacho L, Vollertsen J (2026) Microplastic and tyre wear particles at a highway: a case study from Norway. Microplast Nanoplast 6(1):14

Giechaskiel B, Grigoratos T, Mathissen M, Quik J, Tromp P, Gustafsson M, Franco V, Dilara P (2024) Contribution of road vehicle tyre wear to microplastics and ambient air pollution. Sustainability 16(2). https://doi.org/10.3390/su16020522

Horton AA, Walton A, Spurgeon DJ, Lahive E, Svendsen C (2017) Microplastics in freshwater and terrestrial environments: evaluating the current understanding to identify the knowledge gaps and future research priorities. Sci Total Environ 586:127–141. https://doi.org/10.1016/j.scitotenv.2017.01.190

Iordachescu L, Rullander G, Lykkemark J, Dalahmeh S, Vollertsen J (2024) An integrative analysis of microplastics in spider webs and road dust in an urban environment–webbed routes and asphalt trails. J Environ Manage 359. https://doi.org/10.1016/j.jenvman.2024.121064

Kirstein IV, Hensel F, Gomiero A, Iordachescu L, Vianello A, Wittgren HB, Vollertsen J (2021) Drinking plastics?—Quantification and qualification of microplastics in drinking water distribution systems by μFTIR and Py-GCMS. Water Res 188. https://doi.org/10.1016/j.watres.2020.116519

Liu F, Vianello A, Vollertsen J (2019) Retention of microplastics in sediments of urban and highway stormwater retention ponds. Environ Pollut 255. https://doi.org/10.1016/j.envpol.2019.113335

Monira S, Roychand R, Bhuiyan MA, Hai FI, Pramanik BK (2022) Identification, classification and quantification of microplastics in road dust and stormwater. Chemosphere 299. https://doi.org/10.1016/j.chemosphere.2022.134389

Morioka T, Tanaka S, Yamada Y, Yukioka S, Aiba F (2023) Quantification of microplastic by particle size down to 1.1 μm in surface road dust in an urban city, Japan. Environ Pollut 334. https://doi.org/10.1016/j.envpol.2023.122198

Muñiz R, Rahman MS (2025) Microplastics in coastal and marine environments: a critical issue of plastic pollution on marine organisms, seafood contaminations, and human health implications. J Hazard Mater Adv 18. https://doi.org/10.1016/j.hazadv.2025.100663

Rasmussen LA, Liu F, Klemmensen NDR, Lykkemark J, Vollertsen J (2024) Retention of microplastics and tyre wear particles in stormwater ponds. Water Res 248. https://doi.org/10.1016/j.watres.2023.120835

Sahai H, Aguilera del Real AM, Alcayde A, Bueno MJM, Wang C, Hernando MD, Fernández-Alba AR (2025) Key insights into microplastic pollution in agricultural soils: a comprehensive review of worldwide trends, sources, distribution, characteristics and analytical approaches. TrAC Trends Anal Chem 185. https://doi.org/10.1016/j.trac.2025.118176

Shen M, Zeng Z, Wen X, Ren X, Zeng G, Zhang Y, Xiao R (2021) Presence of microplastics in drinking water from freshwater sources: the investigation in Changsha, China. Environ Sci Pollut Res 28(31):42313–42324. https://doi.org/10.1007/s11356-021-13769-x

Sridharan S, Kumar M, Singh L, Bolan NS, Saha M (2021) Microplastics as an emerging source of particulate air pollution: a critical review. J Hazard Mater 418. https://doi.org/10.1016/j.jhazmat.2021.126245

Sunny AR, Sazzad SA, Islam MA, Mithun MH, Hussain M, Raposo A, Bhuiyan MKA (2025) Microplastics in aquatic ecosystems: a global review of distribution, ecotoxicological impacts, and human health risks. Water 17(12). https://doi.org/10.3390/w17121741

Vianello A, Boldrin A, Guerriero P, Moschino V, Rella R, Sturaro A, Da Ros L (2013) Microplastic particles in sediments of Lagoon of Venice, Italy: first observations on occurrence, spatial patterns and identification. Estuar Coast Shelf Sci 130:54–61. https://doi.org/10.1016/j.ecss.2013.03.022

Zúñiga Umaña JM, Paniagua SA, Brenes RC, Vega Baudrit JR (2025) Microplastic pollution in Costa Rican marine ecosystems: origins, ecotoxicological impacts, and mitigation strategies. Mar Policy 179. https://doi.org/10.1016/j.marpol.2025.106772

Chapter 11
Biodegradable Polymer Blends with Lignin as UV-Stabilizer for Eco-Sustainable Fishing Nets

Roberta D'Auria, Immacolata Liotta, Roberto Avolio, Rachele Castaldo, Maria Emanuela Errico, Gennaro Gentile, Federico Olivieri, and Mariacristina Cocca

Introduction

In the last few decades, plastic and microplastic pollution in marine environments has been an increasing global issue, with significant ecological consequences (Jaikumar et al. 2025; Santonicola et al. 2025). It was estimated that 14 million tons (Mt) of plastic end up in the ocean every year (IUCN 2024), part of these wastes consists of abandoned or discarded fishing gear. Abandoned or lost fishing gear, known as "*ghost nets*", is one of the most damaging sources, because it persists to damage habitats, and increases plastic accumulation in the oceans (Liotta et al. 2024). In addition to their impact on the marine ecosystem, ghost nets represented a significant source of microplastic pollution. Once plastic material enters the marine environment, they were exposed to physical, chemical, and biological stressors that promote their progressive degradation and fragmentation. This process led to the formation of microplastics, typically defined as insoluble solid particles smaller than 5 μm, which persist in seawater and pose growing ecological and toxicological risks (Federici et al. 2022). In this context, biodegradable polymers could be a valid alternative for their ability to decompose under environmental conditions through biological processes into products safer and more environmentally benign (Dallaev et al. 2025). In particular, bio-based polymers such as polybutylene succinate (PBS) and polybutylene adipate terephthalate (PBAT) were selected to produce blends as a sustainable alternative to produce fishing nets (Muthuraj et al. 2014).

R. D'Auria (✉) · I. Liotta · R. Avolio · R. Castaldo · M. E. Errico · G. Gentile · F. Olivieri · M. Cocca
Institute for Polymers, Composites and Biomaterials, National Research Council of Italy, Pozzuoli, Italy
e-mail: robertadauria@cnr.it

G. Gentile et al. (eds.), *Proceedings of the 4th International Conference on Microplastic Pollution in the Mediterranean Sea*, Springer Water,
https://doi.org/10.1007/978-3-032-23847-4_11

Lignin (L) a natural UV-absorbing compound (Zhang and Naebe 2021), was introduced into PBS/PBAT polymer blends, to improve the materials' stability when exposed to sunlight in a marine environment, using its natural ability to absorb UV radiation (250–400 nm). In this work, different lignin concentrations in a blend containing PBS (85 wt%) and PBAT (15 wt%) were morphologically analyzed and then colorimetric analysis post UV exposure was performed.

Materials and Methods

The materials were melt-mixed by the means of a Brabender internal mixer, Plastograph EC (Brabender GmbH & Co. KG, Germany) at 180 °C and 60 rpm. 60 g of polymers and different amounts of lignin powder, according to Table 11.1, were added to the mixture. The extrudate was then cooled at room temperature and cut into smaller pieces to obtain uniform samples suitable for further testing. The different lignin concentrations used as filler in the blend are reported in Table 11.1.

The blends were subsequently characterized through melt flow index (MFI) measurements and viscosity analysis using a rotational rheometer. Specifically, MFI testing was conducted at 180 °C under a 2.16 kg load, using a Ceast Melt Flow Junior 6943 melt flow index tester. Approximately 5 g of pelletized blend were loaded into the melt flow indexer. The instrument, through its integrated software, calculates the MFI based on the weight of the extrudate collected over a specific time interval. Viscosity measurements were then conducted using the rotational rheometer (Thermo Scientific Haake RheoStress 6000), with parallel plate geometry (25 mm diameter, 1 mm gap), under controlled temperature at 180 °C and shear conditions. The shear rate was varied from 0.5 to 300 s^{-1}, and the resulting viscosity values were recorded for each sample. Data analysis was carried out by extracting viscosity values at different shear rates using the rheometer's software. The compounded materials were further processed into polymeric films by compression molding using a hydraulic press (Collin P200E, Ebersberg, Germany) at 180 °C, 4 min at 0 bar and 1 min at 20 bar, obtaining 0.15 μm thick films. Polymeric films were exposed to both natural and artificial UV radiation (using Osram Ultra-Vitalux 300W) to evaluate yellowing over time. Colorimetric measurements, at different intervals, were performed using Spectrophotometer CM-23d (Konica Minolta, Inc. Japan) to assess

Table 11.1 Blend, PHR lignin, and code used for the blends

Lignin (phr)	Code
0	PBS-PBAT
0.1	0.1 L
0.5	0.5 L
1	1 L
2.5	2.5 L

the yellowing of the different films. Morphological analysis was performed on the top surface of the films using a FEI Quanta 200 Field-Emission Gun (FEG) Scanning Electron Microscope (SEM) (FEI, Hillsboro, OR, USA) working in high vacuum mode with an acceleration voltage ranging from 10 to 30 kV and using a secondary electron detector. Before the analysis, samples were mounted on aluminum stubs and coated with Au/Pd alloy by means of an Emitech K575X sputtering device.

Results and Discussion

SEM micrographs (Fig. 11.1a, b) collected before UV exposure (days 0) showed a smooth and homogeneous surface for the PBS/PBAT blend. While in the filled samples, it was clearly visible the presence of lignin that appeared as a small irregularity on the top and fracture surface, making them rougher.

The results of MFI and viscosity tests showed an increase in blend MFI and a decrease of the blend viscosity by increasing the lignin content. Colorimetric analysis was performed after 75 days of exposure to UV radiation.

Blends characterized by lower lignin contents tended to show reduced yellowing.

These results may suggest that moderate lignin loadings could be associated with more effective UV protection. Conversely, samples containing higher lignin concentrations appeared to exhibit higher ΔE * ab values, which indicate that further increases in filler content do not necessarily translate into additional improvements in UV protection. This behavior could reflect a complex, non-linear interaction between lignin concentration and photo-induced color changes within the polymer matrix.

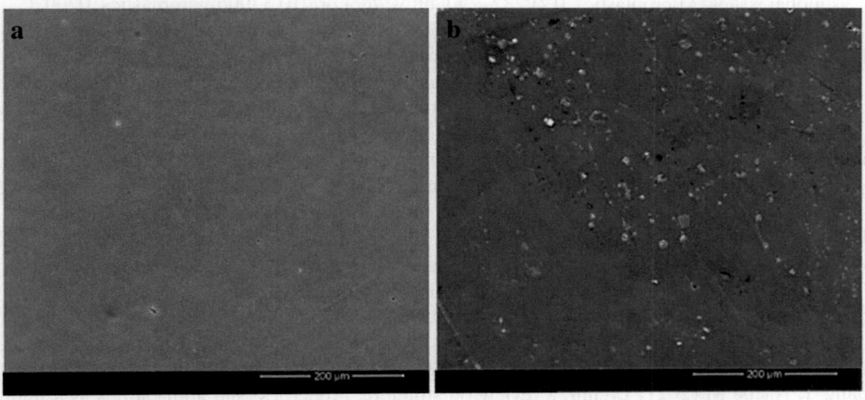

Fig. 11.1 SEM micrographs of the top section of PBS/PBAT 85:15 and the sample with filler

Conclusions

In this work, PBS-PBAT based blends containing different lignin contents were examined to obtain preliminary information on their behavior and stability under UV exposure. Colorimetric analysis was performed over a period of 75 days, which allowed an initial evaluation of photo-induced yellowing in the materials considered. The results suggested that blends with relatively lower lignin contents tend to exhibit reduced color variation, confirmed by lower $\Delta E*ab$ values, while higher filler concentrations did not appear to offer additional benefits in terms of UV protection within the investigated conditions. Future works will aim to extend the colorimetric monitoring over longer UV exposure times and different ageing conditions. In addition, post-exposure morphological analyses will be performed to assess possible degradation phenomena induced by UV light. Thermal analyses are also planned to compare the properties of the materials in the initial state and after UV exposure, thus providing a more comprehensive evaluation of the effect of yellowing on the investigated blends.

References

Dallaev R, Papež N, Allaham MM, Holcman V (2025) Biodegradable polymers: properties, applications, and environmental impact. Polymers 17(14). https://doi.org/10.3390/polym17141981

Federici S, Ademovic Z, Amorim MJB, Bigalke M, Cocca M, Depero LE, Dutta J, Fritzsche W, Hartmann NB, Kalčikova G, Keller N, Meisel TC, Mitrano DM, Morrison L, Raquez JM, Tubić A, Velimirovic M (2022) COST action PRIORITY: an EU perspective on micro- and nanoplastics as global issues. Microplastics 1(2):282–290. https://doi.org/10.3390/microplastics1020020

IUCN © (2024) IUCN website IUCN issues briefs: Twitter: @IUCN. www.iucn.org/issues-briefs

Jaikumar IM, Tomson M, Meyyazhagan A, Balamuralikrishnan B, Baskaran R, Pappuswamy M, Kamyab H, Khalili E, Farajnezhad M (2025) A comprehensive review of microplastic pollution in freshwater and marine environments. Green Anal Chem 12:100202. https://doi.org/10.1016/J.GREEAC.2024.100202

Liotta I, Avolio R, Castaldo R, Gentile G, Ambrogi V, Errico ME, Cocca M (2024) Mitigation approach of plastic and microplastic pollution through recycling of fishing nets at the end of life. Process Saf Environ Prot 182:1143–1152. https://doi.org/10.1016/j.psep.2023.12.031

Muthuraj R, Misra M, Mohanty AK (2014) Biodegradable poly(butylene succinate) and poly(butylene adipate-co-terephthalate) blends: reactive extrusion and performance evaluation. J Polym Environ 22. https://doi.org/10.1007/s10924-013-0636-5

Santonicola S, Volgare M, Olivieri F, Cocca M, Colavita G (2025) Natural and regenerated cellulosic microfibers dominate anthropogenic particles ingested by commercial fish species from the Adriatic Sea. Foods 14(7):1237. https://doi.org/10.3390/foods14071237

Zhang Y, Naebe M (2021) Lignin: a review on structure, properties, and applications as a light-colored UV absorber. ACS Sustain Chem Eng 9(4):1427–1442. https://doi.org/10.1021/acssuschemeng.0c06998

Chapter 12
Towards Sustainable Solutions Based on Biopolymeric Blends for Containment Nets in the Aquaculture Sector

Cristina De Monte, Lucia Ricci, Leonardo Arrighetti, Umberto Granata, Carlo Andrea Massa, Michele Magri, Silvia Merlino, and Simona Bronco

Introduction

Plastic pollution is one of the major environmental concerns affecting marine ecosystems, with an estimated 4.8–12.7 million tons of plastic entering the oceans each year (Wang et al. 2021). Among the many sources of plastic debris, aquaculture contributes through the widespread use of conventional plastic materials. In offshore fish farming, hundreds of plastic cable ties are routinely used to repair damaged nets. Despite careful handling, many of these ties are inevitably lost at sea, where they may persist for decades, contribute to microplastic generation, or even be ingested by fish.

The development of biodegradable cable ties offers a promising alternative to mitigate this issue. The challenge is to identify polymers that can degrade under real marine conditions without releasing harmful substances, while maintaining sufficient durability during their service life. Compostable materials such as PLA and PBAT have shown limited biodegradation in seawater (Karamanlioglu et al. 2017; Narancic et al. 2018; De Monte et al. 2023a), highlighting the need for innovative

C. De Monte (✉) · L. Ricci · L. Arrighetti · U. Granata · C. A. Massa · S. Bronco
Institute for Chemical and Physical Processes, National Research Council of Italy, Pisa, Italy
e-mail: cristina.demonte@cnr.it

L. Arrighetti
Department of Chemistry and Industrial Chemistry, University of Pisa, Pisa, Italy

M. Magri
Associazione Ambienti Magri, Livorno, Italy

S. Merlino
Institute for Marine Sciences (ISMAR), National Research Council of Italy, Lerici, Italy

G. Gentile et al. (eds.), *Proceedings of the 4th International Conference on Microplastic Pollution in the Mediterranean Sea*, Springer Water,
https://doi.org/10.1007/978-3-032-23847-4_12

approaches. Recent research has explored polyhydroxyalkanoates (PHAs), biocomposites reinforced with natural fibers, and PBS/PBAT blends, all of which demonstrate improved degradation profiles in marine environments (Strangis et al. 2023; Le Gué et al. 2023). In addition to biodegradability, antifouling properties are crucial. Conventional antifouling strategies rely on biocides, raising concerns about toxicity and regulatory acceptance. Sustainable alternatives, including bioinspired coatings, hybrid polymers, and natural compounds such as capsaicin, are under investigation (Li et al. 2023). Against this backdrop, the ECOSEA project focuses on the development of bio-based, seawater-degradable, antifouling cable ties for aquaculture. This innovation aims to reduce plastic pollution, improve fish welfare, and foster a more sustainable and environmentally responsible aquaculture industry.

Materials and Methods

Biodegradable polymer pellets as polylactic acid, PLA (Ingeo 2003D®, NatureWorks, Plymouth, MN, USA), polybutylene adipate-co-terephthalate, PBAT (Ecoflex® F Blend C1200, BASF, Ludwigshafen, Germany), Polybutylene succinate, PBS (PBE001, NaturePlast, Mondeville, France), polybutylene succinate-co-adipate, PBSA (PBE003, NaturePlast, Mondeville, France) and p-hydroxy-benzoate hydroxylase polyhydroxybutyrate, PHBHB (IamNature B6 G713 grade, MAIP, Settimo Torinese, Italy) were used. To develop the formulations, additives derived from waste biomass such as fibrous waste with a prevalent lignocellulosic nature (labeled as BF and WF) or polyester materials with a high degree of hydroxylation (labeled PI), supplied by Stazione Sperimentale per l'Industria delle Conserve Alimentari-SSICA (Parma, Italy) and by TomaPaint (Canneto sull'Oglio, Mantova, Italy), were selected. Additives were maintained frozen, as received and was lyophilized, at—50 °C and 0.01 mbar (Pascal-LIO 5PDGT) for two days before using (Arrighetti et al. 2024). The flakes and fibers were dried in a vacuum oven at 60 °C for 5 h before being processed (De Monte et al. 2023b) The reactive processing was performed in the discontinuous mixer (Brabender Plastograph EC), with a mixing chamber of 30 cm^3, interfaced with Brabender Mixing Software "Win Mix 1.0" for the data management and acquisition and in a single-screw extruder (Brabender Measuring Extruder 19/25 D connected to Plastograph Can) equipped with a compounding screw (Barrier Screw, Maillefer, compression ratio: 2.5:1). The rotor speed was set to 50 rpm, for 10 min at 180 °C (PLA-PBAT and PLA-PHB blends) or 150 °C (PBS-PHB blend), in the case of discontinuous mixer, and for 2 min applying the following temperature profile: 200, 200, 200, 200 °C, in the case of single-crew extruder. Blends were pre-dried at 60 °C overnight in vacuum conditions to remove humidity absorbed and prototypes were produced by Direct3D Printer F30 Pro (Direct3D, Milano, Italy), equipped with Direct3D Pellet Extruder, with heated bed in the temperature range 90–290 °C (with Marlin 2.0 Firmware).

Results and Discussion

Novel material formulations were developed via melt compounding of biodegradable polymers with biomolecules derived from food and industrial byproducts. This processing route enabled the production of tailored feedstocks for additive manufacturing, which were subsequently used to fabricate 3D-printed prototypes. The resulting specimens underwent comprehensive performance evaluation and accelerated marine aging tests to assess their mechanical integrity and degradation behaviour under operational conditions.

Specimens Formulations for Cable-Tie Prototypes and Sea Trials

The formulations selected to produce the prototypes consist of pure polymers, binary polymer blends or blends of polymers and lignocellulosic additives (BF, WF) or highly hydroxylated polyester additive (PI) (Table 12.1).

A Cable tie prototype specimen (useful length: 180 mm, width: 5 mm, thickness: 1 mm) were designed with the help of FreeCAD software and the 3D models were sliced using free slicing software. In the 3D printing process, prototypes were obtained by adjusting instrumental parameters such as layer thickness, nozzle diameter, print plate temperature, printing speed and nozzle temperature allowed to obtain the prototypes. Figure 12.1 describes 3D printing process and specimens produced.

Sea trials of the prototypes (September 2025) with anchoring structures deployed at different depths (Fig. 12.2).

Table 12.1 Specimens

	PHBHB	PLA	PBAT	PBS	PBSA	BF	WF	PI
BIO-A		X	X					
BIO-B	X	X						
BIO-C					X			
BIO-D				X				
BIO-E	X							
BIO-BF1	X			X		X		
BIO-BF2			X	X		X		
BIO-WF1		X	X				X	
BIO-PI					X			X

Fig. 12.1 3D printing (center) with some specimens produced (right and left)

Fig. 12.2 Sea trials

Conclusions

The ECOSEA project demonstrates the feasibility of redesigning cable ties for aquaculture facilities by developing fastening systems specifically suited to marine environments. By integrating biodegradable biopolymers and biomolecules derived from food and industrial byproducts with direct 3D-printed prototyping and field testing under real marine conditions, the project offers a promising pathway toward achieving marine-grade mechanical properties while ensuring controlled biodegradation and enhanced environmental sustainability. Its dual objective is to deliver products that guarantee reliable mechanical performance during use and to ensure biodegradation within timeframes compatible with environmental protection.

Acknowledgements The authors thank PR FESR Toscana 2021–2027—Action: 1.1.4. Research and development for companies also in groupings with research organizations—Call 2 "Research and development projects for MSMEs and Midcaps", Project "Integrative solutions for aquaculture and ecosistemi health", acronym "ECOSEA", CUP-ST 27717.29122023.043000919, CUP CIPE D57H24003150009. Further thanks go to Dr. Piergiorgio Stipa (Azienda Agroittica Piombino, GR, Italy) and Dr. Simone Tempesti (Porto Turistico Marine Resort of Pisa, Italy) for their valuable support during the tests in the marine environment.

References

Arrighetti L, Ricci L et al (2024) Innovative materials based on physical melt-blending of cutin from tomato waste and poly(lactic acid). Mater Today Sustain. https://doi.org/10.1016/j.mtsust.2024.100852

De Monte C et al (2023a) Resin pellets aging and degradation investigation from long term in situ experiment: first results. In: Cocca M, Ambrogi V, Avolio R, Castaldo R, Errico ME, Gentile G (eds) Proceedings of the 3rd international conference on microplastic pollution in the Mediterranean Sea. ICMPMS 2022. Springer Water. Springer, Cham. https://doi.org/10.1007/978-3-031-34455-8_31

De Monte C et al (2023b) Agro-waste bean fibers as reinforce materials for polycaprolactone composites. Compounds. https://doi.org/10.3390/compounds3030036

Karamanlioglu M et al (2017) Abiotic and biotic environmental degradation of the bioplastic polymer poly(lactic acid): a review. Polym Degrad Stab. https://doi.org/10.1016/j.polymdegradstab.2017.01.009

Le Gué L et al (2023) Mitigating plastic pollution at sea: natural seawater degradation of a sustainable PBS/PBAT marine rope. Mar Pollut Bull. https://doi.org/10.1016/j.marpolbul.2023.115216

Li Z et al (2023) Bioinspired marine antifouling coatings: antifouling mechanisms, design strategies and application feasibility studies. Eur Polym J. https://doi.org/10.1016/j.eurpolymj.2023.111997

Narancic T et al (2018) Biodegradable plastic blends create new possibilities for end-of-life management of plastics but they are not a panacea for plastic pollution. Environ Sci Technol. https://doi.org/10.1021/acs.est.8b02963.

Strangis G et al (2023) Seawater biodegradable poly(butylene succinate-co-adipate)—wheat bran biocomposites. Materials. https://doi.org/10.3390/ma16072593

Wang G-X et al (2021) Seawater-degradable polymers—fighting the marine plastic pollution. Adv Sci. https://doi.org/10.1002/advs.202001121

Chapter 13
Microplastics in Sea Salts from Mediterranean and Atlantic Ocean Salt Pans: A Comparative Study

Cristina Grazia De Nido, Diego Centonze, Carmen Palermo, and Javier Hernández Borges

Introduction

In recent decades, the widespread use and improper management of plastics have led to an environmental crisis, with microplastics and nanoplastics now present in all ecosystems and associated with significant ecological and health risks (Bowley et al. 2021). The durability and low cost of plastics have promoted their use but also accounted for their persistence in the environment, where they accumulate in both terrestrial and aquatic ecosystems. In this context, it is important to emphasize that anthropogenic microparticles found in nature do not all have the same shape, and they should be scientifically distinguished as spheres, fibers, fragments, and films. These categories may originate from natural (e.g. cotton), synthetic (i.e. microplastics), or semi-synthetic (e.g. rayon) materials. This distinction is crucial to fully understand the dynamics of distribution, persistence, and ecotoxicological impact of these anthropogenic particles in ecosystems. Microplastics can originate either from intentional production (primary microplastics, such as cosmetic microbeads, industrial pellets, and glitter) or from the degradation of larger plastic objects (secondary

C. G. De Nido (✉) · D. Centonze
Department of Medical and Surgical Sciences University of Foggia, Foggia, Italy
e-mail: cristina.denido@unifg.it

C. Palermo
Department of Clinical and Experimental Medicine, University of Foggia, Foggia, Italy

J. H. Borges
Departmeto de Química, Área de Química Analítica, Sección de Química, Universidad de La Laguna, Tenerife, Islas Canarias, Spain

Instituto de Enfermedades Tropicales y Salud Pública de Canarias, Universidad de La Laguna, Tenerife, Islas Canarias, Spain

G. Gentile et al. (eds.), *Proceedings of the 4th International Conference on Microplastic Pollution in the Mediterranean Sea*, Springer Water,
https://doi.org/10.1007/978-3-032-23847-4_13

microplastics), a process accelerated by UV radiation, abrasion, oxidation, and temperature fluctuations (Tirkey and Upadhyay 2021). Micro and nanoplastics are an increasing health concern, as they can migrate across different environmental matrices, bioaccumulate in organisms, and reach human tissues through ingestion, inhalation, and dermal contact (Martín-Gómez et al. 2024). Their ubiquity and ability to transport toxic additives justify their classification as high-priority emerging contaminants, and research on this topic is rapidly expanding (Rainieri and Barranco 2019). Microplastic contamination has also been detected in commercial sea salt, raising concerns about potential chronic dietary exposure, particularly given that global production exceeds 300 million tons per year and consumption of this product is widespread (Yang et al. 2015). In this study, we quantified and characterized anthropogenic microparticles in both raw and commercially available salt samples collected from the Mediterranean Sea and Atlantic Ocean salt pans. Characterization included assessment of particle colour, size, shape, and chemical composition. This study contributes to a better understanding of microplastic contamination in food-grade salt and underscores the importance of standardized contamination-free analytical protocols in environmental and food safety research.

Materials and Methods

Sea salt samples from different salt pans were selected from two geographic areas: the Mediterranean Sea and the Atlantic Ocean. For each sample, 50 g of salt were weighed in a glass beaker and then dissolved in 20 μm pre-filtered Milli-Q water and mechanically shaken for 30 m. The solution was then filtered through 50 μm and 25 μm stainless-steel filters that were dried overnight at 50 °C. Filters were visualized with a Leica MZ16FA stereomicroscope, and each anthropogenic microparticles detected was counted, photographed, and evaluated for shape, color, and maximum size. All analyses were carried out in an ISO 6 clean room, using plastic-free equipment washed with filtered water. Glassware was heated at 550 °C for 4 h to reduce contamination. Filtrations were performed under a glove box in the ISO 6 clean room, and procedural blanks were processed every three samples to check the absence of contamination. The characterization of anthropogenic microparticles was performed using a micro-Fourier transform infrared spectrometer (μFTIR) from Thermo Scientific™ (Nicolet™ iN10) in reflection mode. μFTIR analysis was performed on ten microparticles taken from the three replicates, being at least 10% of the overall number, to ensure a reliable and statistically meaningful characterization of the analyzed batch. The resulting spectra were compared with several spectral libraries. Matches $\geq$ 70% were considered valid for the identification, while matches between 60% and 70% were manually assessed.

Results and Discussion

As shown in Table 13.1, the batches show a total number of anthropogenic microparticles detected in the range 52–82 and a number per gram in the range 0.35–0.55. According to the statistical analysis, the MED-2 batch resulted significantly different and the most contaminated (highest number of microparticles) with a mean value of 0.55 ± 0.01 items/g. The overall results suggest that the corresponding geographical area is particularly exposed to inputs of contaminants of both natural and anthropogenic origin, and that the phenomenon is highly remarkable. MED-3 raw salt ranks second in terms of contamination level (0.47 ± 0.11 items/g) showing a great variability among replicates (range 0.38–0.60 items/g), probably due to its inhomogeneity not being submitted to purification. Batches ATL-1, ATL-2, and MED-1 show lower or comparable values in terms of abundance (0.35–0.37 items/g), with a wider variability in particle size, particularly in ATL-1 (up to 3978 µm). This could indicate a more heterogeneous input of anthropogenic microparticles, likely linked to multiple sources and more complex dispersion processes. Figure 13.1 shows the percentage distribution of the different types of anthropogenic microparticles (fibers, films, fragments, and tangled mess) detected in the samples analyzed. In all sampling sites, fibers constitute the predominant category, representing nearly the entirety of the observed particles. Films are notably present in MED-1, MED-2, and ATL-1, while fragments are detected only in ATL-2 at a very low percentage. Overall, it can be highlighted that fibers are the most found in all the samples, and that the percentage of films is higher in the batches from the Mediterranean Sea. As can be seen in Fig. 13.2, depicting the chemical composition of the anthropogenic microparticles, cellulose is the most abundant in all the samples, followed by other cellulosic fibers of natural (i.e. cotton) or semi-synthetic (i.e. rayon) origin. It is noteworthy that cellulosic materials do not represent a significant risk to human health because, in contrast to microplastics, they are biodegradable and tend to have a lower ecotoxicological impact. Conversely, polyester (including PET) is of particular concern due to both its high risk for human health and its huge release from synthetic textiles, industrial discharges, or urban wastewater. Indeed, these types of microplastics are known for their persistence in the environment and their potential interactions with marine organisms, thus representing a possible ecotoxicological risk. In some samples, particularly MED-1 and ATL-2, there are a considerable number of unidentified particles, for which the match quality factor was below the threshold of 70% used for reliable polymer identification. However, a manual spectral assessment shows that many of these microparticles have similarities with semi-synthetic materials, suggesting that their inclusion in the "unidentified" category is more likely due to analytical limitations than to a true inability to characterize them. To summarize, in all the samples, microplastics were found in low percentages (MED-1 33%; ATL-2: 20%; MED-3 raw salt: 10%) while the most detected anthropogenic microparticles were of natural or semi-synthetic origin (in the range 67%–100%). Moreover, the protocol of investigation used showed no significant external contamination, evidenced by the absence of particles in three

Table 13.1 Anthropogenic microparticles detected in Mediterranean Sea and Atlantic Ocean salt pans ATL-1/ATL-2: commercial salt from Atlantic 1–2 salt pans; MED-1/MED-2: commercial salt from Mediterranean 1–2 salt pans; MED-3 raw salt: raw salt from Mediterranean 3 salt pans

Sample	Total number of anthropogenic microparticles found (n = 3)	Items/g ± SD	Anthropogenic microparticles size range (μm)
ATL-1	52	0.35 ± 0.06	190–3978
ATL-2	54	0.36 ± 0.13	103–3076
MED-1	55	0.37 ± 0.04	61–1790
MED-2	82	0.55 ± 0.01	119–2378
MED-3 raw salt	71	0.47 ± 0.11	237–4927

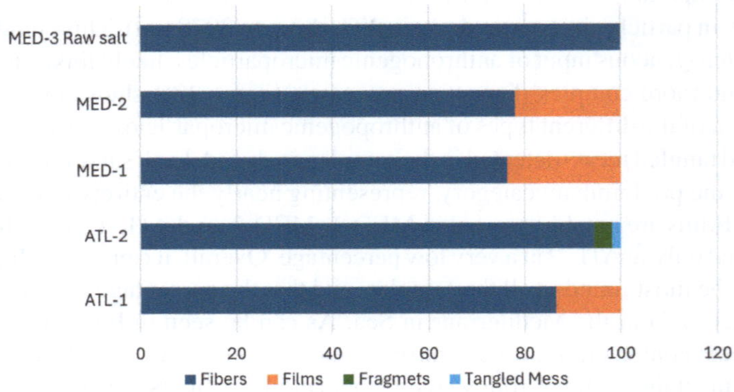

Fig. 13.1 Anthropogenic microparticle shape distribution in the batches analyzed. Each horizontal bar shows coloured proportional segments for each kind of shape reported in the legend at the bottom. ATL-1/ATL-2: commercial salt from Atlantic 1–2 salt pans; MED-1/MED-2: commercial salt from Mediterranean 1–2 salt pans; MED-3 raw salt: raw salt from Mediterranean 3 salt pans

procedural blanks and the presence of only one microparticle each in the other two blank replicates.

Conclusions

This research highlighted the widespread presence of anthropogenic microparticles in sea salt from the Mediterranean Sea and Atlantic Ocean salt pans. A predominance of fibers, mainly of natural or semi-synthetic origin, was observed. Batches from the Mediterranean area, particularly MED-2, showed the highest levels of contamination, but also a remarkable homogeneity among replicates, suggesting a constant input of anthropogenic microparticles. The presence of polyester, although in low percentages, confirms a potential ecotoxicological risk due to the environmental persistence

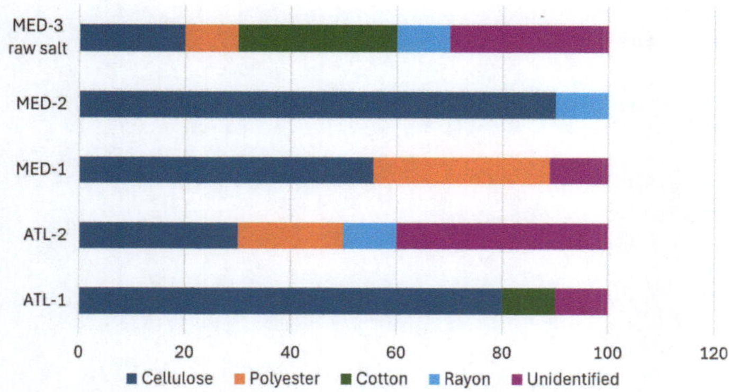

Fig. 13.2 Anthropogenic microparticle composition distribution in the batches analyzed. Each horizontal bar shows coloured proportional segments for each category reported in the legend at the bottom. ATL-1/ATL-2: commercial salt from Atlantic 1–2 salt pans; MED-1/MED-2: commercial salt from Mediterranean 1–2 salt pans; MED-3 raw salt: raw salt from Mediterranean 3 salt pans

of microplastics. The non-identification of anthropogenic microparticles is primarily related to instrumental limitations (match factors around 50%), since the identification output was unambiguous. Among all anthropogenic microparticles detected, the presence of microplastics was low, and the contamination was largely dominated by cellulosic fibers (either natural or semi-synthetic). This study contributes to a better understanding of microplastic contamination in food-grade salt and highlights the importance of standardized protocols to avoid external contaminations, so ensuring data reliability for comparative studies. These findings also emphasize the need for continued monitoring of anthropogenic microparticles to assess the potential of chronic exposure and risks to human health.

References

Bowley AJ et al (2021) Oceanic hitchhikers assessing pathogen risks from marine microplastic. Trends Microbiol 29(2):107–116

Martín-Gómez B et al (2024) Recent applications of chromatography for determining microplastics and related compounds (bisphenols and phthalate esters) in food. Microchem J 197:109903

Rainieri S, Barranco A (2019) Microplastics, a food safety issue? Trends Food Sci Technol 84:55–57

Tirkey A, Upadhyay LSB (2021) Microplastics: an overview on separation, identification and characterization of microplastics. Mar Pollut Bull 170:112604

Yang D et al (2015) Microplastic pollution in table salts from China. Environ Sci Technol 49(22):13622–13627

Chapter 14
Five Years Monitoring Plastics Contamination in Fresh Water of Lake Garda with Seabin. Seasonal Effects, Microplastics and Alien Species

Luca Fambri, Marco Papparotto, Jeevan Kishore Reddy Pidapa, Alessio Fonnesu, Paolo Matteotti, and Claudia Gavazza

Introduction

Microplastic (MP) pollution (Thompson et al. 2004) is becoming an increasingly alarming issue, both in seawater and in freshwater (Azevedo-Santos et al. 2021). The Seabin (https://seabin.io/home) is a convenient mitigation technique for plastic pollution (Nikiema and Asiedu 2022); it is a floating container with a 2 mm-mesh net that filters the surface of calm water bodies (25 m^3 per hour), trapping floating debris, easy to use and low energy consumption (about 0.5 kW). In this work we summarize five years of monitoring and characterization not only of plastic and microplastic, but also alien species with the aid of two Seabins in northern Lake Garda, with a comparative evaluation of seasonal effects.

L. Fambri (✉) · M. Papparotto · J. K. R. Pidapa · C. Gavazza
Department of Industrial Engineering and INSTM Research Unit, University of Trento, Trento, Italy
e-mail: luca.fambri@unitn.it

A. Fonnesu
Istituto Tecnico Tecnologico "G. Marconi", Rovereto, TN, Italy

P. Matteotti
Fraglia della Vela Riva, Riva del Garda, TN, Italy

G. Gentile et al. (eds.), *Proceedings of the 4th International Conference on Microplastic Pollution in the Mediterranean Sea*, Springer Water,
https://doi.org/10.1007/978-3-032-23847-4_14

Materials and Methods

Two Seabins V5 were installed in two ports in the municipality of Riva del Garda (Trento, Italy) in northernmost side of Lake Garda: San Nicolò in 2019 (SN; coordinates 45.877763, 10.857128) and Fraglia della Vela Riva in 2021 (FV; coordinates 45.883455, 10.843897). They are used to collect environmental samples of floating debris, and are then periodically emptied into municipal waste.

The Seabins were turned on and left to work for various days (1–5), they automatically collected the samples with an estimated flow of 25 m^3 of water per hour. Once full, the bins were weighed with a luggage scale, emptied, and the content was dried in environmental conditions. Between 5 and 40 g of floating material is collected per hour in dependence on season. Collected material was manually sorted, separating plant and animal organisms from plastic and other floating waste. The weight percentage of plastics in the range 1–5% in Summer, and about 0.5% in Winter. Plants and animals were weighed and identified. Plastic components were separated in macroplastics and microplastics and sorted into visually recognizable product categories, counted and weighed separately. The plastic particles were then singularly analyzed via FTIR (PerkinElmer Spectrum Two™ portable FTIR spectrometer).

Results and Discussion

The relative and qualitative composition of the plastic waste collected in selected various Seasons from July 2021 to Spring 2025 is showed in Fig. 14.1, detailing the total number of pieces and the main constituents: i.e. polyethylene (PE), polypropylene (PP), expanded polystyrene (EPS), and "other" polymeric materials, such as polyethylene terephthalate (PET), polylactic acid (PLA), polyamides (PAs). The main representative data collected in the first five years during Summer, which is the season with the highest plastic concentration, are reported in Table 14.1.

The main constituent is consistently PE, followed by PP, EPS, and "other", except for sampling during April 2025 in SN, which is an area with high foams residue. An analysis of the seasonal variation of macroplastic in the lake shows a fluctuation in number and an increase in mass concentration from Summer 2021 to Summer 2025 (see Table 14.1). Furthermore, the mass concentration of microplastics decreased significantly from 2021 onwards, before becoming rather variable over time. Results vary across seasons and years, but there is a clear decrease in the number of MP collected per hour from 2021 to 2025 (Fig. 14.2) with an average weight of individual microplastics ranging from 2 to 8 mg.

A list of alien plant and animal species collected with the Seabins in Spring 2025 is reported in Table 14.2 and shown in Fig. 14.3. An average of 0.5 individuals/hour is observed in Seabin FV and 1.3 individuals/hour for Seabin SN, during the considered season. Reported data highlight a decrease in plastic and microplastic collection during the Autumn and Winter seasons with respect to Spring and Summer.

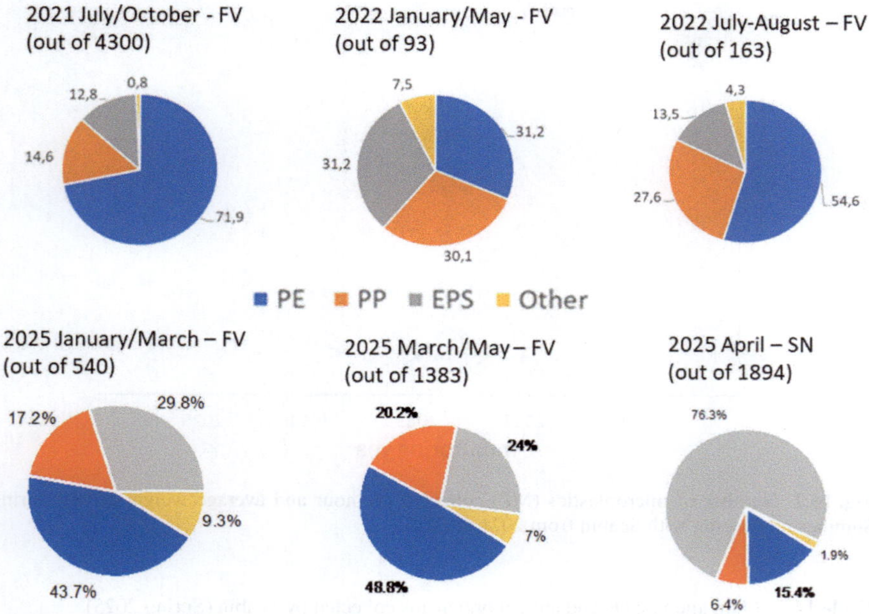

Fig. 14.1 Relative composition by number of plastic items (either macro—and micro-plastics) during various seasons collected in Fraglia della Vela (FV) and in port San Nicolò (SN)

Table 14.1 Values of macro—and microplastic concentration (in terms of mass and number of items collected) during the Summers collected in Fraglia Vela (FV)

Season	Total working time (h)	Macroplastic number concentration (n/m^3)	Microplastic number concentration (n/m^3)	Macroplastic mass concentration (g/m^3)	Microplastic mass concentration (mg/m^3)
Summer 2021	269	0.023	0.667	0.007	2.41
Summer 2022	96	0.033	0.084	0.010	0.68
Summer 2023	24	0.078	0.152	0.007	0.28
Summer 2024	215	0.103	0.060	0.042	0.14
Summer 2025	168	0.037	0.125	0.456	0.45

This could be related to increased anthropic pressure on the lake during the warmer seasons, as Lake Garda is a popular Summer touristic attraction (Papparotto et al. 2025). The presence of some alien species (already previously reported in the area (Ciutti et al. 2011) was confirmed with the employed methodology.

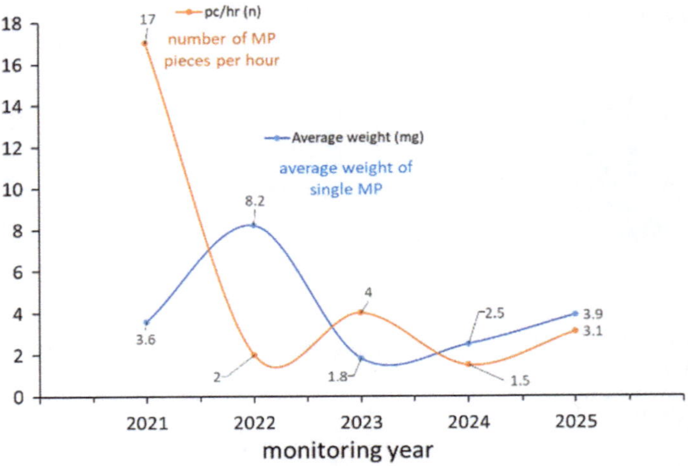

Fig. 14.2 Number of microplastics (MP) collected per hour and average weight of MP during Summer monitoring with Seabin from 2021 to 2025

Table 14.2 Main alien plants and animal organisms collected by Seabin (Spring 2025)

Kingdom	Habitat	Class	Taxa	Location*
Plantae	Aquatic	Alismatales	*Lagarosiphon major*	F/SN
Plantae	Aquatic	Ceratophyllales	*Ceratophylum demersus*	F/SN
Animalia	Aquatic	Crustacea	*Dikerogammarus villosus*	F/SN
Animalia	Aquatic	Bivalvia	*Dreissena bugensis*	F
Animalia	Aquatic	Bivalvia	*Dreissena polymorpha*	F/SN
Animalia	Terrestrial	Insecta	*Limnoxenus* sp.	F
Animalia	Terrestrial	Insecta	*Halyomorpha halys*	SN

*F (Fraglia); SN (San Nicolò)

Conclusions

In this communication, we presented some results from the first five years (2021–2025) of a long-term study on various environmental stressors (plastic pollution and alien species presence) in northern Lake Garda. To this end, we showed that the simple and low-cost floating plastic pollution mitigation technology of the Seabin can be effectively used as alternative environmental sampling technique. Observations indicate an increase in concentrations of floating plastic during Spring and Summer, mainly composed of PE, PP, and EPS. At least six alien species, already reported, were also observed. Seabin monitoring activities have been gradually extended to local high school classes and to promoting awareness and education not only during

Fig. 14.3 *Lagarosiphon major* **a**; *Ceratophylum demersus* **b**; *Dikerogammarus villosus* **c**; *Dreissena* sp. **d**

national and international sailing competitions (https://diinews.unitn.it/en/microplastics-a-macro-problem), but also during Clean-UP Days (You are Patron—https://www.patron-nature.org/en/cleanup).

Acknowledgements The authors thank LifeGate and COOP Italia for their promotion and support activities. The authors thank Vanna Benini (Gruppo Sommozzatori Riva del Garda, https://www.grupposommozzatoririva.it/it) for identifying the aquatic plant taxa. Moreover, Riva del Garda Fierecongressi and ITAS Mutua are gratefully acknowledged for providing the portable Perkin Elmer Spectrum Two™ spectrometer. This research was partially founded by PRIN 2022: PLASTACTS-Assessment of nano/microplastics impacts (Grant No. 202293AX2L, CUP E53D23008370006).

References

Azevedo-Santos V et al (2021) Plastic pollution: a focus on freshwater biodiversity. Ambio 50:1313–1324. https://doi.org/10.1007/s13280-020-01496-5

Ciutti F et al (2011) Non-indigenous invertebrates, fish and macrophytes in Lake Garda (Italy). J Limnol 70:315–320. https://doi.org/10.4081/jlimnol.2011.315

Nikiema J, Asiedu Z (2022) A review of the cost and effectiveness of solutions to address plastic pollution. Environ Sci Pollut Res 29:24547–24573. https://doi.org/10.1007/s11356-021-18038-5

Papparotto M et al (2025) Seasonal comparative monitoring of plastic and microplastic pollution in Lake Garda (Italy) using seabin during summer–autumn 2024. Microplastics 4:44–68. https://doi.org/10.3390/microplastics4030044

Thompson RC et al (2004) Lost at sea: where is all the plastic? Science 304:838. https://doi.org/10.1126/science.1094559

Chapter 15
Recent Developments on Microplastics Analysis Marine Sediments and Air Involving GCMS-Pyrolysis and Agilent LDIR Spectroscopy

Camilla Fantasia, Federico Sacco, and Paolo Sscardina

Introduction

The widespread use of synthetic polymers has led to a global environmental crisis, with microplastics found in many environments and posing risks to human health and ecosystems. Microplastics (MPs) are emerging contaminants that require advanced analytical approaches. For a complete assessment, both morphological characterization and mass quantification are necessary (Pomata et al. 2024; Soll et al. 2025). This study proposes a multi-technique approach to analyze MPs in marine sediments and airborne particulate matter, combining quantitative pyrolysis-GC/MS with infrared chemical imaging using the Agilent LDIR system (Subharthe and Bradley 2020, 2024). The work also highlights the use of innovative tools for sample preparation. Sample preparation is the most critical step, as it must remove interfering substances while preserving the integrity of microplastic particles. One of the main goals of this work was to develop a single preparation method suitable for both analytical techniques (Soll et al. 2025).

C. Fantasia (✉) · F. Sacco
SRA Instruments, Cernusco Sul Naviglio (MI), Italy
e-mail: fantasia@srainstruments.com

P. Sscardina
Agilent Tecnologies, Cernusco Sul Naviglio (MI), Italy

G. Gentile et al. (eds.), *Proceedings of the 4th International Conference on Microplastic Pollution in the Mediterranean Sea*, Springer Water,
https://doi.org/10.1007/978-3-032-23847-4_15

Materials and Methods

Sample Preparation Optimization: A Single Workflow for LDIR and Pyrolysis

Although analyzing microplastics is a complex task, the biggest challenge is often preparing the sample correctly (Soll et al. 2025). Modern spectroscopy and chromatography instruments are highly sensitive, but they require clean samples. Raw environmental materials cannot be introduced directly into a pyrolyzer or an infrared spectrometer (Subharthe and Bradley 2020, 2024).

This makes it necessary to follow a careful and structured preparation procedure that selectively removes everything that is not microplastic. The process usually involves steps such as homogenizing the sample, removing organic matter, separating plastic from non-plastic materials based on density, and filtering to collect the particles.

The exact procedure can vary depending on the sample type—whether sediment, water, or air—but the goal remains the same: to produce samples that can be reliably analyzed by instruments.

In this study, special attention was given to the sample prep phase to eliminate matrix interferences and ensure proper homogenization. All procedures were carried out under ultra-clean conditions to avoid secondary contamination. This approach allowed for reliable and reproducible results, showing how essential proper preparation is in microplastic analysis (Soll et al. 2025).

Innovative Tools for Improving Sample Preparation

Future improvements may include tools like the ECO-F cups from Frontierlab, which allow direct filtration of liquid samples into the same cup used for pyrolysis. This avoids sample transfer steps and reduces time, loss, and contamination.

Another useful tool is the IQ MILL-2090 Bio, a fast and powerful mill that helps break down complex materials. It allows the simultaneous grinding of multiple samples and improves the homogeneity and representativeness of solid materials.

Although these tools were not used in this study, they offer promising ways to make environmental sample preparation faster and more efficient.

Integrated Analytical Approach

After proper preparation, samples were analyzed using two complementary techniques, providing complete qualitative and quantitative information.

The Agilent LDIR 8700 (Laser Direct Infrared) allows automated morphological analysis and polymer identification in about one hour (Subharthe and Bradley 2020).

The Frontier Lab pyrolyzer, connected to an Agilent GC/MS system, provides accurate mass quantification of polymers (Subharthe and Bradley 2024).

Frontier Lab 3030D Pyrolyzer with Agilent GC/MS

Thermal analysis and analytical pyrolysis break down polymers using heat and analyze the resulting fragments. These are powerful methods for studying polymeric materials such as microplastics.

In this study, the Frontier Lab Multi-Shot Pyrolyzer (EGA/PY-3030D) was used to perform both pyrolysis and thermal desorption.

Quantification was done using pyrolysis-GC/MS in splitless mode. Calibration curves for 10 common polymers were created using the Microplastics Calibration Standard from Frontier Lab, with SiO_2 as the solid diluent.

The F-Search MPs software helped generate the curves quickly and automatically, allowing for sensitive and selective quantification of polymers down to nanogram levels. This made the process faster and more reliable (Subharthe and Bradley 2024).

LDIR Technology with Quantum Cascade Laser (QCL)

The LDIR 8700 system is an infrared imaging instrument that uses a quantum cascade laser (QCL), fast-scanning optics, and a thermoelectrically cooled MCT detector. It overcomes many limitations of traditional FTIR systems and allows fast measurement of particle size, shape, and chemical composition (Subharthe and Bradley 2020).

It automatically detects all particles in a defined area, traces their edges, takes an image, and compares their spectrum to a reference library to identify the polymer type.

Filtration is essential to capture suspended microplastics, but standard filters can release fibers or fail to retain small particles.

Agilent provides metal-coated filters (gold or aluminum) with 0.8 μm pore size, compatible with vacuum filtration and directly usable in the LDIR system.

Alternatively, IR low-e glass slides can be used to deposit microplastics when dealing with a higher number of particles. Filters are preferred when particle concentration is low, while slides are better for dense samples.

This method simplifies the workflow and ensures accurate and fast results.

In this study, microplastics from marine sediments were cleaned, filtered, and analyzed automatically with the LDIR system. The instrument scanned the filters, located the particles, measured their shape and size, and identified the polymer type using its spectral library.

Results and Discussion

Results—Py-GC/MS

From the qualitative and quantitative analysis of airborne particles collected on Teflon filters, a complex chemical profile was observed, with styrene, typical of polystyrene, as the main peak.

The most common polymers identified were polypropylene, polyethylene, ABS, PMMA, polycarbonate, nylon 6, nylon 6,6, PVC, and PET.

Concentrations ranged from 3.0 to 190 μg/g, with polystyrene being the most abundant, followed by PVC and PET.

Results—LDIR Analysis

The analysis found microplastic particles larger than 20 μm in concentrations between 10 and 226 particles/m^3.

A total of 30,471 natural particles and 635 synthetic particles were detected.

The most common polymer types were acrylics/polyurethanes/paints (39.2%), PET (26.0%), PE-CI (7.1%), PVC (6.0%), PE (5.2%), PP (5.2%), and rubber (4.3%).

About 94.9% of all particles had a diameter smaller than 300 μm.

Conclusions

This study demonstrates the effectiveness of a combined analytical approach for the detection and quantification of microplastics in both marine sediments and airborne particulate matter. The integration of Py-GC/MS and Agilent LDIR allowed for complementary qualitative and quantitative analysis, ensuring reliable polymer identification and concentration measurement (Subharthe and Bradley 2020, 2024). Special emphasis on sample preparation proved crucial to avoid interferences and achieve reproducible results. Although not yet applied, innovative tools such as ECO-F cups and IQ MILL-2090 show promising potential for improving sample handling efficiency (Soll et al. 2025). The results highlight the presence of a wide range of polymer types, with polystyrene, PET, and PVC being the most abundant. The LDIR system enabled fast and automated morphological analysis, revealing that most particles are smaller than 300 μm. This workflow confirms the importance of multi-technique strategies in environmental microplastics research and opens the door for future standardization in monitoring protocols.

References

Pomata D et al (2024) Integrated approaches for the characterization of microplastics in environmental matrices. Sci Total Environ (Elsevier) 932:17303

Soll M, Watanabe A, Pipkin W (2025) Advances in microplastics analysis. In: Frontier lab laboratory whitepaper, frontier laboratories Ltd.

Subharthe S, Bradley C (2020) Microplastics analysis using Agilent LDIR 8700: a new infrared chemical imaging approach. In: Agilent application note 5994–2421 (Italian edition), Agilent Technologies

Subharthe S, Bradley C (2024) Quantitative analysis of polymer mass using pyrolysis-GC/MS with F-Search MPs software. In: Agilent application note 5994–7587 (English edition), Agilent Technologies

Chapter 16
Sustainable-To-Go Cups: Eco-Friendly or Hidden Plastic Hazard Under Thermal Stress?

Berenice Dafne García García, Gurusamy Kutralam-Muniasamy, Fermín Pérez-Guevara, and V. C. Shruti

Introduction

Human exposure to plastic particles is a growing concern, particularly via oral intake, which remains the primary route of entry into the body (Li 2023). Ingestion has been associated with oxidative stress, inflammatory lesions, tissue translocation, and chronic conditions (Prata 2020), indicating that dietary plastics may pose an emerging health risk. Plastic particles have been detected in bottled water, beer, salt, tea bags, and seafood (Smith 2018). A significant source of exposure stems from single-use containers; disposable cups, commonly used for hot beverages, release considerable amounts of particles during use, especially at elevated temperatures (Chen 2023). Traditionally lined with polyethylene (PE) or polystyrene (PS), these cups are increasingly being replaced by "sustainable" alternatives such as polylactic acid (PLA). However, the performance of these plastics under hydrothermal conditions typical of daily use—such as contact with hot liquids—remains inadequately studied, potentially undermining both their functionality and assumed safety. This study, a component of my doctoral research on 22 cups, provides a preliminary comparative analysis of three materials: PE, PS, and PLA. The aim was to evaluate particle release under standardized migration conditions and assess surface

B. D. García García · F. Pérez-Guevara · V. C. Shruti (✉)
Department of Biotechnology and Bioengineering, Centro de Investigación y de Estudios Avanzados del Instituto Politécnico Nacional (CINVESTAV), Ciudad de México, México
e-mail: shruti.venkata@cinvestav.mx

G. Kutralam-Muniasamy
CIITEC-IPN, Centro de Investigación E Innovación Tecnológica, Ciudad de México, CDMX, México

G. Gentile et al. (eds.), *Proceedings of the 4th International Conference on Microplastic Pollution in the Mediterranean Sea*, Springer Water,
https://doi.org/10.1007/978-3-032-23847-4_16

alterations using Attenuated Total Reflectance–Fourier Transform Infrared Spectroscopy (ATR-FTIR), Scanning Electron Microscopy (SEM), flow cytometry, and X-ray Photoelectron Spectroscopy (XPS).

Materials and Methods

The study was structured into three main stages. In the first stage, the cups were analysed using (ATR-FTIR) to confirm their polymeric composition, (SEM) to examine surface morphology, and (XPS) to determine surface chemical composition. These baseline analyses served as a reference for subsequent comparisons.

In the second stage, the cups were subjected to migration tests using deionized water as a food simulant, at 95 °C for 15 min, following European Union regulations (EU 2018).

In the third stage, the particles released into the simulant were quantified by flow cytometry with Nile Red staining, and their morphology was assessed by SEM. Finally, the recovered plastic films were characterized post-migration by SEM and XPS to identify the surface changes induced by the migration process.

Results and Discussion

FTIR-ATR

ATR-FTIR analysis confirmed the polymeric composition of the cup coatings: PLA (C = O stretching at 1745 cm^{-1}, 85%), PE (C–H stretching between 2915 and 2848 cm^{-1}, 97%), and PS (aromatic C = C stretching at 1600 cm^{-1} with characteristic bands at 760–800 cm^{-1}, 95%). These spectral assignments are consistent with the known structural features of the respective polymers and were validated using the OpenSpecy library and previous literature (Smith 2021a, b; Smith 2022) (Fig. 16.1).

Surface Morphology Alterations Post-Migration

Comparison of the coating surfaces before and after the migration tests (Fig. 16.2) revealed distinct morphological changes under SEM. PLA transitioned from a smooth, homogeneous surface to one marked by fractures and cracks, accompanied by the release of amorphous aggregates. PE developed surface roughness and exhibited angular fragment detachment, while PS showed prominent fissures and released both spherical and angular particles, indicative of mechanical degradation. These observations align with previous findings: hydrothermal degradation of PLA

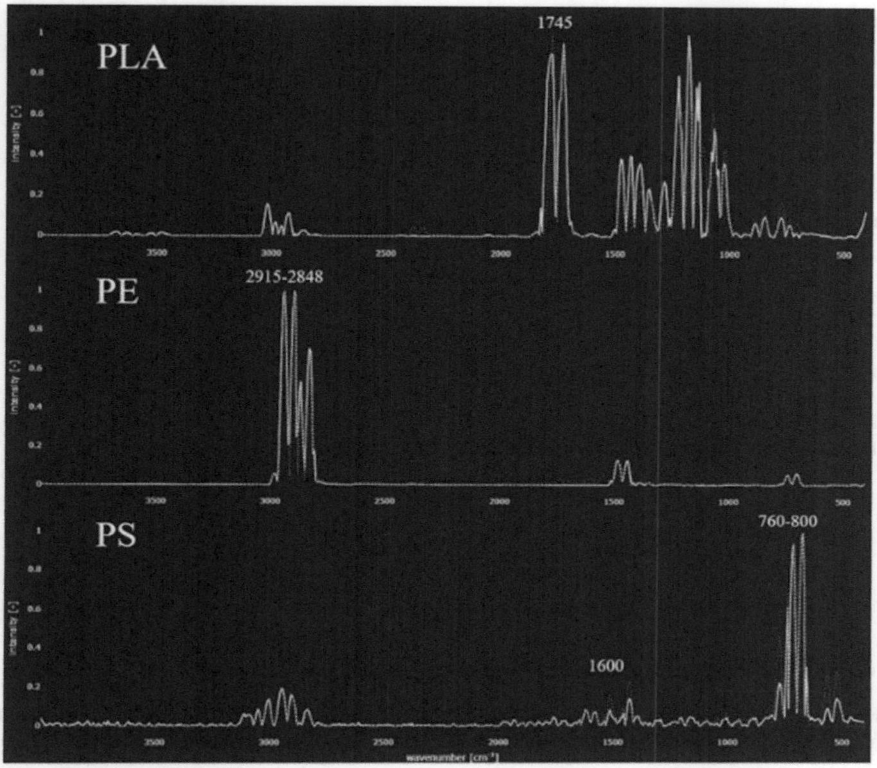

Fig. 16.1 ATR-FTIR spectra of the selected cups: **a** PLA, **b** PE, and **c** PS. Experimental spectra (white) are overlaid with library reference spectra (red) obtained from OpenSpecy

typically results in brittle surfaces and the release of irregular particles (Fang 2024), whereas conventional plastics such as PE and PS tend to develop surface roughness, fissures, and particle detachment under thermal stress (Ranjan 2021).

Flow Cytometric Profiling of Migration-Derived Particles

Flow cytometry enabled the quantification and characterization of particles released from each polymer following migration testing (Fig. 16.3). PLA exhibited more dispersed particle populations, indicating a broader range of sizes and optical properties. In contrast, PE generated compact and relatively homogeneous populations, while PS showed a mixed profile—comprising abundant high-density populations alongside smaller groups with lower optical complexity. Quantitatively, PLA released approximately 10^6 particles/mL, significantly higher than PE and PS, which both released around 10^5 particles/mL. These values substantially exceed those reported

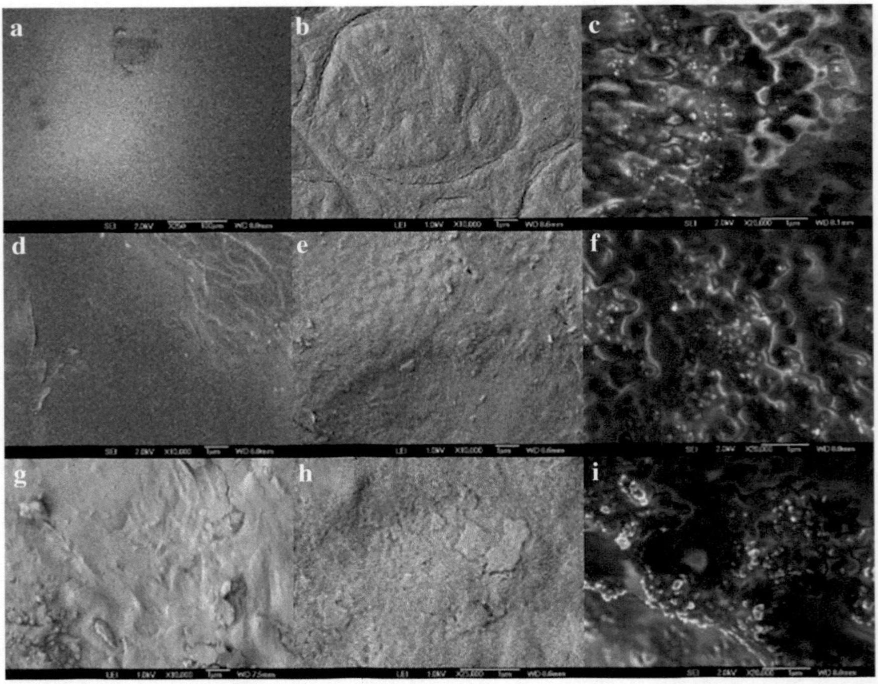

Fig. 16.2 SEM micrographs of the selected cups: plastic film before migration (left), after migration (center), and particles released into the simulant (right). (**a–c**) PLA, (**d–f**) PE, (**g–i**) PS

in previous studies (~1,293 particles/mL for PE (Zhang 2023) and ~ 180 particles/mL for PLA (Wang 2023), highlighting the impact of testing conditions and reinforcing the need for standardized protocols in particle release assessments.

Based on these concentrations, the estimated intake per single use reached ~ 250 million particles for PLA, compared with ~ 25 million for PE/PS, highlighting the potential relevance of this exposure route.

Surface Chemical Changes Revealed by XPS

XPS analysis revealed distinct surface degradation mechanisms depending on the polymer type. In PE and PS, increased signals corresponding to C–O and O–C = O bonds were detected, indicative of oxidation processes triggered by hydrothermal exposure. In contrast, PLA exhibited additional peaks associated with ester bond hydrolysis, reflecting its susceptibility to degradation under moist, high-temperature conditions. These chemical alterations are consistent with SEM observations, which revealed fractures and increased surface roughness, as well as with flow cytometry results, which showed a significantly higher release of particles from PLA. Although

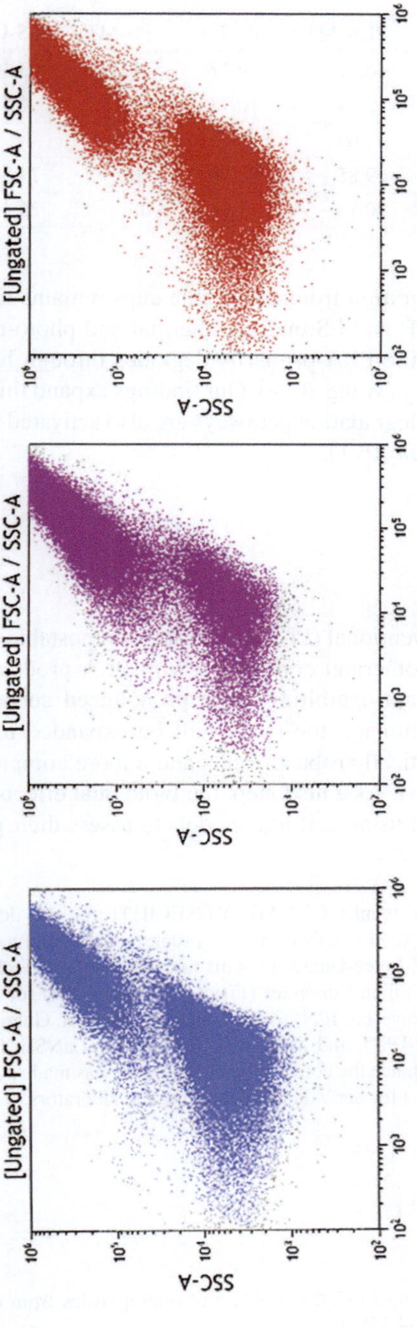

Fig. 16.3 Dot plots (SSC-A versus FSC-A) obtained by flow cytometry with Nile Red staining. **a** PLA, **b** PE, and **c** PS, showing the particle populations released after the migration tests

Table 16.1 Relative areas of chemical bonds obtained by XPS in PLA, PE, and PS. UT: untreated; MT: post-migration (%)

Binding type	PLA-UT	PLA-MT	PE-UT	PE-MT	PS-UT	PS-MT
C–C/C–H	35.49	60.17	89.27	81.05	93.46	77.0
C–O	32.54	22.74	10.73	13.15	6.54	15.7
O–C = O/C = O	31.96	17.09	–	5.80	–	7.30
O1s C–O	47.70	69.85	88.76	100	74.54	95.30
O1s O–C = O	52.30	30.15	11.24	0	25.46	4.70

the literature on particle migration from disposable cups remains limited, previous studies have reported that PE and PS undergo thermal and photo-oxidative degradation (Di Giulio 2024), while PLA primarily degrades through hydrolysis in the presence of heat and humidity (Wang 2024). Our findings expand this understanding by demonstrating that these degradation pathways are also activated under simulated food-contact conditions (Table 16.1).

Conclusions

Disposable cups—both conventional (PE and PS) and compostable (PLA)—release plastic particles under hydrothermal conditions, with PLA producing an order of magnitude more particles and exhibiting more pronounced surface degradation. While this analysis is preliminary, the study will be expanded to include all 22 cup samples to ensure statistically robust results and a more comprehensive understanding of particle migration. As a next step, the biological effects of the released particles will be investigated using cell line models to assess their potential impact on human health.

Acknowledgements The author thanks CONAHCYT/SECIHTI for the doctoral scholarship (CVU: 1146073) and the *Elisa Acuña* (CINVESTAV) program for additional support. Special thanks are also extended to Dr. F. Pérez-Guevara for his financial support. Technical support was generously provided by Dr. A. Guillen-Cervantes (Física, Cinvestav-IPN) for SEM imaging, Dr. S.A. Tomás-Velázquez (Física, Cinvestav-IPN) for XPS analysis, Eng. M. Guerrero-Cruz for FTIR measurements (Física, Cinvestav-IPN), and MSc. V.H. Rosales-G. (LaNSE, Cinvestav-IPN) for flow cytometry, with whose assistance the completion of this work was made possible. The author also thanks MSc. M. Pucheta-Díaz for her valuable support in the laboratory.

References

Chen H, Xu L, Yu K, Wei F, Zhang M (2023) Release of microplastics from disposable cups in daily use. Sci Total Environ 854:158606

Di Giulio T, De Benedetto GE, Ditaranto N, Malitesta C, Mazzotta E (2024) Insights into plastic degradation processes in marine environment by X-ray photoelectron spectroscopy study. Int J Mol Sci 25:5060

Fang C, Zhang Z, Zhang X, Naidu R (2024) Microplastics or micro-bioplastics released by wrinkling paper cup. Sci Total Environ 945:174123

Li Y, Tao L, Wang Q, Wang F, Li G, Song M (2023) Potential health impact of microplastics: a review of environmental distribution, human exposure, and toxic effects. Environ Health 1:249–257

Prata JC, da Costa JP, Lopes I, Duarte AC, Rocha-Santos T (2020) Environmental exposure to microplastics: an overview on possible human health effects. Sci Total Environ 702:134455

Ranjan VP, Joseph A, Goel S (2021) Microplastics and other harmful substances released from disposable paper cups into hot water. J Hazard Mater 404:124118

Smith BC (2021a) The infrared spectra of polymers III: hydrocarbon polymers. Spectroscopy 36(11):22–25

Smith B (2021b) The infrared spectra of polymers II: polyethylene. Spectroscopy 36(9):24–29

Smith BC (2022) Infrared spectroscopy of polymers, VIII: polyesters and the rule of three. Spectroscopy 37(10):25–28

Smith M, Love DC, Rochman CM, Neff RA (2018) Microplastics in seafood and the implications for human health. Curr Environ Health Rep 5:375–386

Union E (2018) Commission regulation (EU) 2020/1245. Off J Eur Union 2016:48

Wang JJ, Yang L, Li D, Shi Y, Hill C, Pilliadugula R, Xiao L (2023) High levels of microparticles release from biodegradable polylactic acid paper cups compared with polyethylene-lined cups. Chem Eng J 468:143620

Wang Y et al (2024) Comparing the aging processes of PLA and PE: the impact of UV irradiation and water. Processes 12:635

Zhang T, Hu JL, Duan Y, Chen S, Li D, Dong B, Lin QB (2023) Identification and characterisation of microplastics released from plastic-coated paper cups using micro-Raman spectroscopy. Food Control 153:109901

Chapter 17
Design of Hyper-crosslinked Cellulose Nanofibrils as Renewable Sorbents for Chlorophenol Removal from Water

Antonio Maglione, Antonia Cerbone, Federico Olivieri, Roberto Avolio, Rachele Castaldo, Mariacristina Cocca, Maria Emanuela Errico, Veronica Ambrogi, and Gennaro Gentile

Introduction

Water contamination by aromatic pollutants, such as chlorinated phenols, represents a serious environmental and health concern due to their toxicity, persistence, and potential carcinogenicity. These pollutants originate from industrial effluents including agrochemical, pharmaceutical, and pesticide manufacturing, and are often resistant to biodegradation, making their removal from water challenging.

Adsorption is widely regarded as an effective, versatile, and cost-efficient technology for water purification because it does not produce toxic by-products, can be spontaneous and reversible, and allows adsorbent regeneration (Burevska-Atkovska et al. 2024; Castaldo et al. 2023; Guerritore et al. 2022). Traditional adsorbents such as activated carbon and synthetic porous polymers exhibit high performance but are often derived from fossil fuels and may lack sustainability. Cellulose, the main

A. Maglione · A. Cerbone · F. Olivieri · R. Avolio · R. Castaldo · M. Cocca · M. E. Errico · G. Gentile (✉)
Institute for Polymers, Composites and Biomaterials National Research Council of Italy, Pozzuoli, NA, Italy
e-mail: gennaro.gentile@cnr.it

R. Castaldo
e-mail: rachele.castaldo@cnr.it

A. Cerbone
Department of Chemical Sciences, University of Naples Federico II, Naples, Italy

V. Ambrogi
Department of Chemical, Materials and Industrial Production Engineering, University of Naples Federico II, Napoli, Italy

G. Gentile et al. (eds.), *Proceedings of the 4th International Conference on Microplastic Pollution in the Mediterranean Sea*, Springer Water,
https://doi.org/10.1007/978-3-032-23847-4_17

constituent of lignocellulosic biomass, offers significant promise as a sustainable and biodegradable resource for a wide range of industrial uses (Pirozzi et al. 2023).

In this study, cellulose nanofibrils (CNF) were functionalized with an aromatic polymer phase and subjected to a hyper-crosslinking by Friedel–Crafts alkylation reaction to produce a micro/mesoporous material with high specific surface area, aiming to combine high adsorption efficiency for aromatic pollutants with a large biobased content.

The target pollutant in this work was 2,4-dichlorophenol (DCP), a halogenated aromatic compound frequently used as a model contaminant to evaluate adsorbent performance in aqueous systems (Maglione et al. 2025).

Materials and Methods

In this study, cellulose nanofibrils (CNF) were used for the synthesis of a highly porous adsorbent. The nanofibrils were first chemically modified to introduce reactive carboxyl groups on their surface, enabling subsequent grafting of poly(vinylbenzyl chloride) (PVBC) chains. These grafted chains served as a precursor for a hyper-crosslinked polymer network, which was created via a Friedel–Crafts alkylation reaction. The reaction conditions, including reagent ratios, temperature, and time, were carefully optimized to maximize the specific surface area and the synthetic yield.

The resulting hyper-crosslinked cellulose nanofibrils (xCNF-PVBC) were characterized using a combination of physicochemical techniques: Fourier-transform infrared spectroscopy, solid state ^{13}C cross-polarization (CP) magic angle spinning (MAS) NMR, thermogravimetric analysis (TGA). Nitrogen adsorption–desorption measurements allowed determination of the specific surface area and pore size distribution. The morphology of the fibrils and the cross-linked polymer phase was examined by scanning and transmission electron microscopy.

The adsorption performance was evaluated using aqueous solutions of 2,4-dichlorophenol (DCP), a representative aromatic pollutant. Adsorption experiments were conducted at controlled temperature and neutral pH, with the adsorbent exposed to varying concentrations of DCP until equilibrium was reached. The effect of competing species, such as natural organic matter represented by humic acids, was also investigated to mimic realistic environmental conditions. Adsorption capacities and removal efficiencies were calculated from concentration measurements before and after equilibration.

Results and Discussion

The hyper-crosslinked cellulose nanofibrils xCNF-PVBC exhibited a remarkable increase in surface area and porosity compared to unmodified cellulose. Nitrogen adsorption analysis indicated that the material developed a high density of micropores while maintaining a significant fraction of mesopores, providing a network of accessible adsorption sites suitable for capturing aromatic molecules. Electron microscopy images revealed that the fibrillar structure was preserved despite the extensive polymer crosslinking, demonstrating that the synthesis produced a material with both high surface area and structural integrity. Spectroscopic analyses confirmed the successful grafting of the polymer and the formation of a crosslinked network, which contributed to the hydrophobic and aromatic character favorable for adsorbing chlorinated phenols.

Adsorption experiments demonstrated that the hyper-crosslinked nanofibrils were highly effective at removing DCP from water. At moderate initial concentrations, the material achieved near-complete removal, and even at higher pollutant loadings, the adsorption capacity reached impressive values of several hundred milligrams per gram. The adsorption process was spontaneous and exothermic, consistent with strong interactions between the aromatic pollutant and the polymer-modified cellulose surface. The presence of humic acids slightly reduced the adsorption efficiency, but the material retained most of its performance, highlighting its potential applicability in natural waters containing organic matter.

The adsorbent also exhibited excellent stability and reusability. Repeated adsorption–desorption cycles revealed minimal loss in capacity, indicating that the hyper-cross-linked cellulose nanofibrils could be regenerated multiple times without structural deterioration. These results suggest that the combination of high surface area, favorable pore architecture, and chemical functionality renders the material highly suitable for practical water remediation applications targeting aromatic contaminants.

Conclusions

This work demonstrates that hyper-crosslinked cellulose nanofibrils can serve as highly efficient and sustainable adsorbents for aromatic pollutants in aqueous environments. The synthetic strategy successfully generated a material with a three-dimensional porous network, high specific surface area, and preserved nanofibrillar morphology, resulting in a strong affinity for chlorinated phenols. The adsorbent displayed remarkable adsorption capacity, rapid uptake, and robust performance even in the presence of competing natural organic matter. Importantly, it retained its effectiveness over multiple adsorption–desorption cycles, highlighting its potential for repeated use in water treatment applications.

Beyond its performance metrics, the material is largely biobased, offering an environmentally friendly alternative to conventional fossil-derived adsorbents. These findings indicate that hyper-crosslinked cellulose nanofibrils could be further developed for large-scale water purification technologies, potentially employing cellulose from agro-industrial residues and biobased aromatic crosslinkers to enhance sustainability. Overall, this study underscores the promise of combining nanocellulose scaffolds with hyper-crosslinked polymer networks to create high-performance, renewable adsorbents for the removal of persistent aromatic contaminants from water.

Acknowledgements The authors acknowledge financial support under the projects: • "Photocatalytically-regenerable hierarchically porous adsorbents for efficient water treatment—PHOTOPAD", code P2022FP2W4, CUP B53D23027540001, PRIN 2022 PNRR financed under the National Recovery and Resilience Plan (NRRP), Mission 4, Component 2, Investment 1.1, Call for tender No. 1409 published on 14.09.2022 with Grant Assignment Decree No. 1389 adopted on 01.09.2023 by the Italian Ministry of University and Research (MUR), funded by the European Union NextGenerationEU. • "HIerarchically Porous materials with tailored ADsorption properties to face environmental pollution—HIPAD", Proposal code FIS-2023–01686, CUP B53C24009520001, financed by the Italian Science Fund (FIS 2 Call) of the Italian Ministry of University and Research (MUR), MUR Grant Assignment Decree No. n. 23314 adopted on 11/12/2024

References

Burevska-Atkovska K, Olivieri F, Avolio R, Castaldo R, Cocca M, Errico ME, Gentile G, Grozdanov A (2024) Amino-modified microporous hyper-crosslinked resins for heavy metal ions adsorption. Colloids Surf A: Physicochemical Eng Asp 700:134720. https://doi.org/10.1016/j.colsurfa.2024.134720

Castaldo R, Berezovska I, Silverstein MS, Gentile G (2023) PolyHIPEs containing hyper-crosslinked resins: hierarchical porosity with broad and versatile sorption properties. ACS Appl Polym Mater 5(9):6920–6931. https://doi.org/10.1021/acsapm.3c00974

Guerritore M, Olivieri F, Avolio R, Castaldo R, Cocca M, Errico ME, Lavorgna M, Silvestri B, Ambrogi V, Gentile G (2022) Hierarchical micro-to-macroporous silica nanoparticles obtained by their grafting with hyper-crosslinked resin. Microporous Mesoporous Mater 335:111864. https://doi.org/10.1016/j.micromeso.2022.111864

Maglione A, Olivieri F, Avolio R, Castaldo R, Cocca M, Errico ME, Ambrogi V, Gentile G (2025) Hyper-cross-linked cellulose nanofibrils with spontaneous and reversible adsorption of aromatic pollutants from water as a valid alternative to fossil-based adsorbents. ACS Appl Mater Interfaces 17(26):38619–38634. https://doi.org/10.1021/acsami.5c05009

Pirozzi A, Olivieri F, Castaldo R, Gentile G, Donsì F (2023) Cellulose isolation from tomato pomace: part II—integrating high-pressure homogenization in a cascade hydrolysis process for the recovery of nanostructured cellulose and bioactive molecules. Foods 12(17):3221. https://doi.org/10.3390/foods12173221

Chapter 18
Environmental Potential Risk of Microplastic Pollution Due to Agricultural Plastics: A Case Study in Apulia Region

Ali Hachem, Fabiana Convertino, Evelia Schettini, and Giuliano Vox

Introduction

The use of plastic products has brought significant benefits to agricultural activities. It has become integral to modern agriculture, serving essential purposes such as boosting productivity by providing protective covering like films, nets and mulching films, protecting crops from pests and diseases, regulating temperature and conserving moisture. Moreover, plastic irrigation systems helped in allowing precise irrigation, minimizing wastage and improving water efficiency (Vox et al. 2016). Other benefits include extending the growing seasons, preserving food safety and quality through packaging and storage solutions.

In contrast, plastic mismanagement led to the production of huge amounts of agricultural plastic waste (APW), contributing to environmental pollution. A major challenge is the effective management of agricultural plastics, particularly at the end of their useful life. A serious and increasingly concerning issue is that these plastics are a potential source of microplastics (MPs) release into the soil (King et al. 2023).

To optimize their performance during production while also addressing the environmental concerns they pose, correct management of agricultural plastics becomes paramount. It is essential to implement effective strategies that mitigate the negative impact of plastic waste on the environment. One actionable approach is to focus on limiting the problem of APW upstream, by reducing the quantity of waste generated in the first place. This step forward is crucial in tackling the escalating footprint of APW, which has been a growing concern over the years (Hachem et al. 2023).

A. Hachem · F. Convertino · E. Schettini (✉) · G. Vox
Department of Soil, Plant and Food Sciences, University of Bari, Bari, Italy
e-mail: evelia.schettini@uniba.it

G. Gentile et al. (eds.), *Proceedings of the 4th International Conference on Microplastic Pollution in the Mediterranean Sea*, Springer Water,
https://doi.org/10.1007/978-3-032-23847-4_18

This study points out environmental potential risks of MPs based on a territorial analysis, supported by a Geographic Information System (GIS).

Materials and Methods

To tackle proper plastic waste management, it is important to begin with the classification of APW and understanding the flow within the agricultural areas.

An agricultural area in the Apulia region (Italy) composed by four municipalities was considered as a case study. By analyzing land-use maps, areas generating plastic waste were identified, and plastic waste indices (PWIs) were applied to different crop types and plastic applications to quantify and map APW generation. The municipalities are Mola di Bari, Conversano, Polignano a Mare and Monopoli, covering an area of 400 km^2. The main crop types grown were identified through land use maps and Corine Land Cover and are shown in Fig. 18.1.

Once the crops are identified, the successive step is the identification of the plastic applications used in each group of crops, such as greenhouse films, mulching films, nets, irrigation pipes, fertilizers bags and pesticidal containers. The characteristics of each plastic application are evaluated as well. In fact, each plastic application has its own PWI depending on the characteristics such as the density of the plastic material, the thickness and the life duration (Hachem et al. 2023).

An example of the PWI for mulching films is given hereafter:

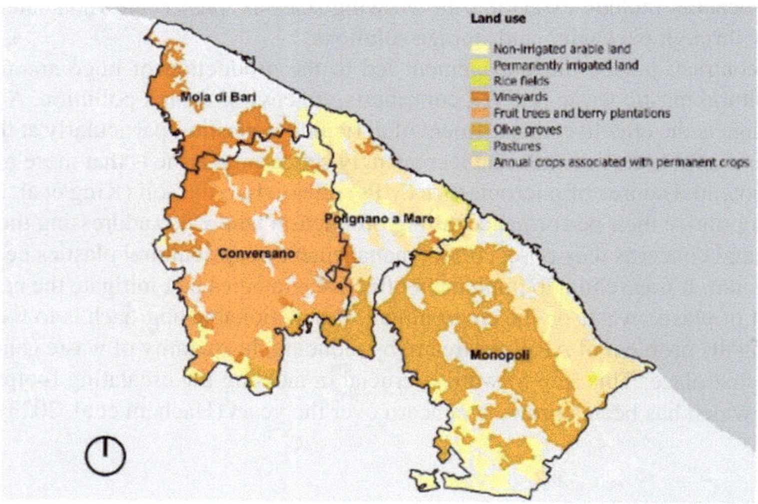

Fig. 18.1 Land use map showing the study area and the crop types grown

$$PWI_{mulchingfilms} = \frac{\rho \cdot T}{years} \cdot F_{correction}$$

where:

- ρ (kg m^{-3}) is the density of the plastic film;
- *T* (mm) is the thickness of the plastic film. The thickness of the plastic material is typically measured in millimeters or microns and is subsequently converted to meters to ensure compatibility with the equation;
- $F_{correction}$ is the dimensionless correction factor used, based on the slope of the plastic material in relation to its coverage on the soil surface;
- *years* is the life duration of the plastic application.

This approach allows for the estimation of the generated plastic waste amount in the specific studied area in kg ha^{-1} year^{-1}.

PWIs combined with land use maps through the GIS allowed for APW georeferencing and having maps of different plastic applications. This is an essential first step for the management of the generated waste.

Further efforts were made to qualitatively assess patterns of MPs contamination from agricultural plastics using risk indices present in literature (FAO 2021; Briassoulis 2023).

The assessment and mapping of MP release in quantitative terms is still a great challenge. However, the evaluation and mapping of the potential risk for MP release from agricultural plastics is a useful exercise, which can guide priority for pollution control actions. To evaluate such potential in relation to the most used agricultural plastic applications, a semi-quantitative metric was developed through the formulation of an index that compares and ranks different geographical areas. This ranking system was called the Agri Plastic Pollution Risk Index. This metric integrates the amount of APW generated from each plastic application in each agricultural area with a corresponding Relative Risk Index (RRI), derived from recent peer-reviewed studies (FAO 2021; Briassoulis 2023) evaluating the likelihood of MPs release into soil per plastic type.

Table 18.1 presents RRIs used to produce the risk map for each plastic application. These risk indices were related to end-of-life scenarios such as agricultural plastic nonrecyclable and/or mismanaged. Mismanagement may be intentional, or it may occur that, during material recovery, accidental losses can take place. This can happen for example for plant supports and clips, fractions of mulching films, nets and irrigation pipes (depending on the thickness and the level of weathering), small fractions of hay bales, and so on.

The risk maps were prepared using the following formula:

$$TotalAPPRI = \sum APW_i \times RRI_i$$

where APW_i is the annual waste generation (in tons per year) from the i—plastic application, and RRI_i is a dimensionless factor (Table 18.1) showing the relative chance of MPs being released into soil from that plastic application. The resulting

Table 18.1 Relative risk index per plastic application

Plastic application	Relative risk index
Containers	11.2
Mulching films	10.5
Irrigation pipes	9.5
Tunnels	8.25
Nets	8
Bags	9.35
Greenhouse films	6.3

APPRI values were calculated for each municipality by adding up all the plastic uses in that area.

Lastly, the data were linked to each municipality in a GIS system, as for the APW maps. The end result was a geo-referenced map showing the semi-quantitative risk of MPs based on APW estimates for each plastic application and municipality.

Results and Discussion

The study found that, while the area is dominated by olive groves, this crop is not the main contributor to APW production. Orchards were identified as the most significant contributors due to the combination of area covered and high PWI from the usage of protection nets and of irrigation pipes (Fig. 18.2a). Orchards account for 60% of the approximately 11,000 tons of APW produced annually in the whole area.

Based on this result and on the RRIs, a qualitative risk map of MPs pollution from agricultural plastics was developed as well (Fig. 18.2b). This shows the range of soil contamination risks from MPs linked to agricultural plastics going from low to high risk. Orchards are the main cause of pollution due to their broad distribution and high risk index associated with the plastic applications used. The municipality of Conversano resulted the municipality with the highest risk in terms of soil potential contamination from MPs due to agricultural plastics.

Conclusions

The study could be useful in drawing attention and promoting effective actions toward the sustainable management of APW and the prevention of MPs contamination. The potential risk maps could contribute to public awareness and perceptions on the impact of MPs due to the use of plastic in agriculture.

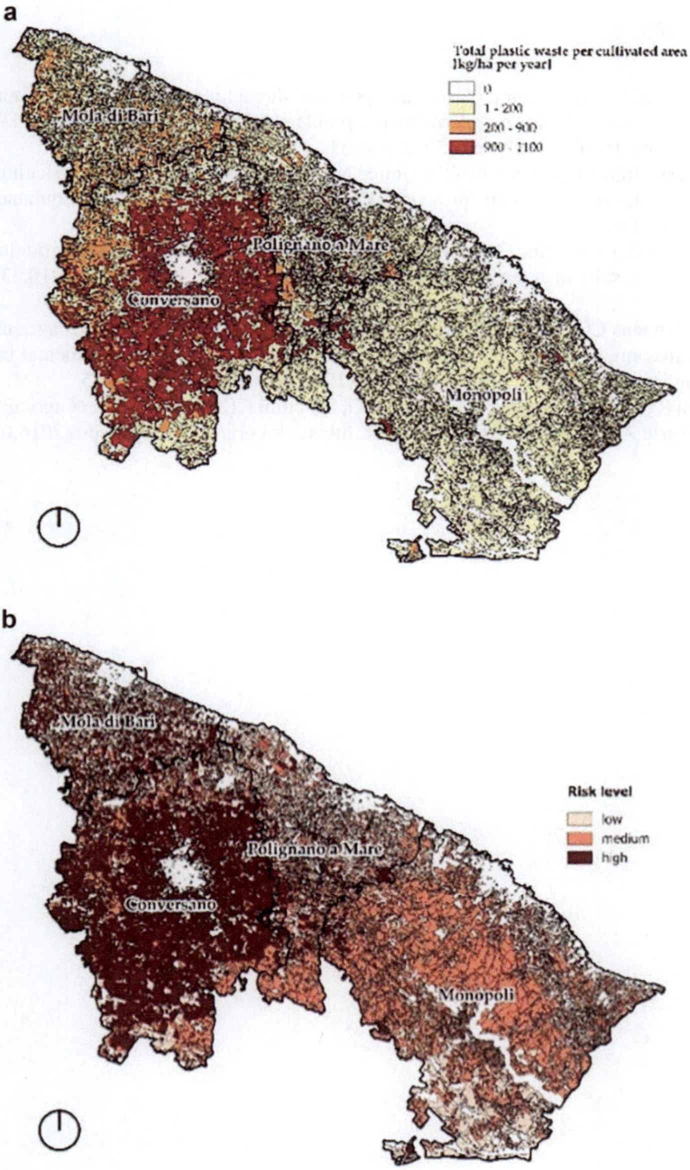

Fig. 18.2 Density of agricultural plastic waste in the study area (**a**) and risk map of microplastics pollution from agricultural plastics (**b**)

References

Briassoulis D (2023) Agricultural plastics as a potential threat to food security, health, and environment through soil pollution by microplastics: problem definition. Sci Total Environ 892:164533. https://doi.org/10.1016/j.scitotenv.2023.164533

Food and Agriculture Organization of the United Nations (2021) Assessment of agricultural plastics and their sustainability: a call for action. FAO. https://openknowledge.fao.org/handle/20.500.14283/cb7856en

Hachem A, Vox G, Convertino F (2023) Prospective scenarios for addressing the agricultural plastic waste issue: results of a territorial analysis. Appl Sci 13:612. https://doi.org/10.3390/app13010612

King CD, Stephens CG, Lynch JP, Jordan SN (2023) Farmers' attitudes towards agricultural plastics—management and disposal, awareness and perceptions of the environmental impacts. Sci Total Environ 864:160955. https://doi.org/10.1016/j.scitotenv.2022.160955

Vox G, Loisi RV, Blanco I, Scarascia Mugnozza G, Schettini E (2016) Mapping of agriculture plastic waste. Agric Agric Sci Procedia 8:583–591. https://doi.org/10.1016/j.aaspro.2016.02.080

Chapter 19
Microwave-Assited Hydrogen Peroxide Degradation of Polystyrene Microplastics

Natalia Hernandez-Montero and Esteban Duran-Herrera

Abstract Polystyrene (PS) is a non-biodegradable polymer whose accumulation poses environmental and health concerns. This study evaluates the degradation of PS microplastics using a microwave-assisted oxidative process with hydrogen peroxide and nitric acid at 220 °C. The particles evolved from "irregular white fragments into yellow-white clusters", confirming progressive structural breakdown. GPC analysis showed the early disappearance of high-molecular-weight chains, while FTIR spectra revealed carbonyl bands near 1700 cm^{-1}, indicating oxidative modification. The treatment achieved an 84 % mass loss and followed a pseudo-first-order kinetic model described as m/m_0=0.976\, e^{-0.015t}. TOC increased substantially, suggesting the release of soluble organic byproducts. These results demonstrate the efficiency of microwave-assisted oxidation for PS degradation and highlight the need to characterize the resulting liquid-phase compounds.

Introduction

Commonly used in packaging and electronics, polystyrene (PS) is a non-biodegradable polymer whose accumulation poses serious environmental and health concerns. Developing efficient and eco-friendly degradation methods is essential. Microwave-assisted degradation has gained attention for its rapid heating and energy efficiency (Xia et al. 2022), though studies on PS degradation in peroxidative-nitric solutions remain limited. Understanding the kinetics of PS degradation under microwave conditions is key to optimizing the process and enabling industrial applications (Schrank et al. 2022). This study investigates the degradation behavior and preliminary kinetics of PS microplastics exposed to microwave irradiation at 220 °C in a peroxidative-nitric solution. The chemical agents promote oxidative breakdown, and the interaction between microwave energy and the peroxidative-nitric solution is

N. Hernandez-Montero (✉) · E. Duran-Herrera
Chemical Engineering Department of the University of Costa Rica (UCR), University of Costa Rica (UCR), San José, Costa Rica
e-mail: natalia.hernandezmontero@ucr.ac.cr

G. Gentile et al. (eds.), *Proceedings of the 4th International Conference on Microplastic Pollution in the Mediterranean Sea*, Springer Water,
https://doi.org/10.1007/978-3-032-23847-4_19

expected to produce a synergistic effect (Mdluli et al. 2024). Ultimately, this research highlights the potential of this technique for developing sustainable microplastic waste treatment strategies.

Materials and Methods

PS particles, sized between 1 and 2.5 mm, were obtained by mechanically fragmenting PS cups and then sieving them through Retsch filters. For the degradation study, the PS samples were dried to a constant weight, and approximately 0.4000 g of PS was weighed. A solution composed of 4 ml of 30% hydrogen peroxide, 1 ml of concentrated nitric acid, and 5 ml of ultrapure MiliQ water was then added to the PS. This mixture was exposed to microwave radiation using a Speed Xpert Berghof D-72800 Eningen microwave oven to investigate its decomposition. The procedure was replicated for time intervals of 0, 15, 30, 40, 50, 60, 75, 90, and 105 min. The recovered PS was dried until constant weight was achieved, and the mass loss data were recorded. Fourier Transform Infrared Spectroscopy (FTIR), Gel Permeation Chromatography (GPC), and morphology analysis were conducted to the solid fraction recovered of PS. The recovered liquid fraction of the treatment was used to investigate the Total Organic Carbon (TOC) composition.

Results and Discussion

Figure 19.1 shows PS SMZ25 stereo zoom microscope from Nikon images and microscopic images using scanning electron microscopy (SEM) Jeol JSM-IT500, before and after the microwave-assisted hydrogen peroxide digestion. After microwave exposure microscopic characterization reveals the transformation of the particles from irregular white fragments into yellow-white clusters. Melted porous clusters morphologies were observed in the first stage of the degradation and clusters of particles were the final morphology. These observations indicate progressive degradation of the PS particles.

The degradation of PS into lower molecular weight compounds and oligomers is evident from mass loss measurements and GPC results. The findings of GPC analysis are illustrated in Fig. 19.2, confirming that the treatment effectively degrades PS into smaller entities. Over time, there is a notable reduction in average molecular weights, indicating significant polymer breakdown. It is suggested that the high-molecular-weight chains are effectively breaking down into oligomers, as it is observed in Fig. 19.2, the first peak, corresponding to high molecular weight chains, disappears at the beginning of the microwave-assisted oxidative process.

An IR spectral study was conducted using a Perkin Elmer FTIR spectrometer and Pike GladiATR to analyze possible degradation bands. FTIR analysis could indicate oxidative modifications in the polymer structure, since bands around 1700 cm^{-1} could

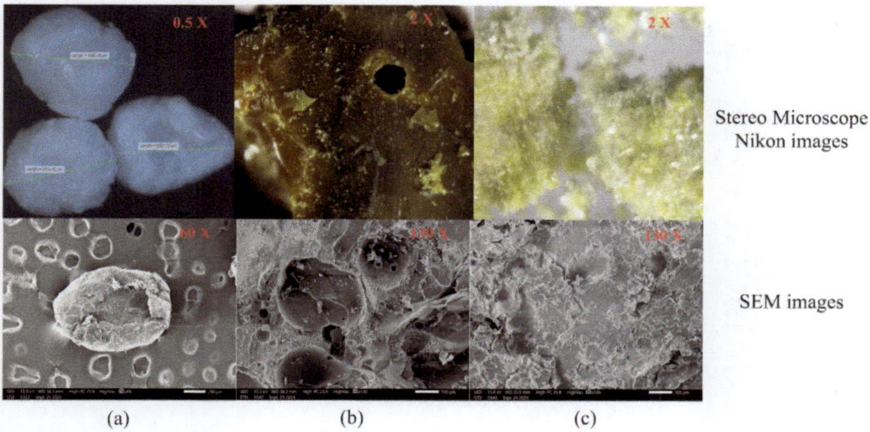

Fig. 19.1 Morphology of PS particles before and after microwave-assisted hydrogen peroxide digestion, as observed through stereo microscope Nikon images and SEM images at different stages: **a** before digestion, size of the particle, **b** at 15 min and, **c** at 105 min

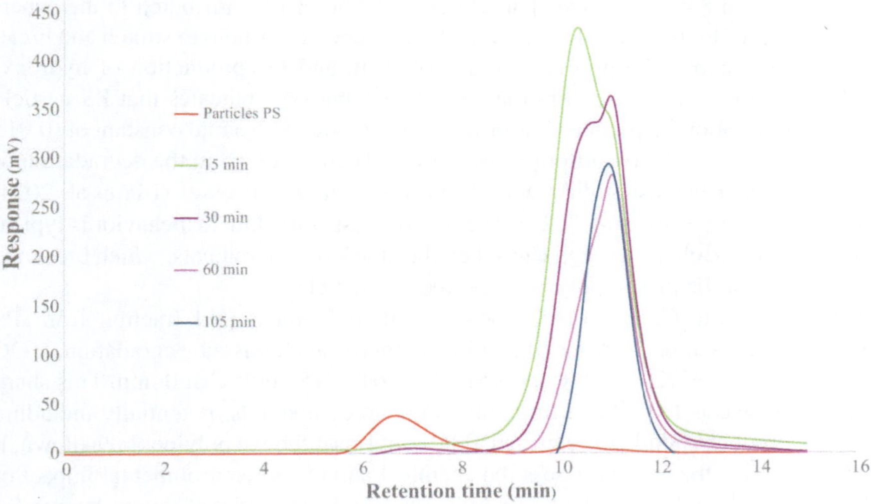

Fig. 19.2 Gel Permeation chromatograms of PS particles in comparison with different times of microwave-assisted hydrogen peroxide digestion

be associate to carbonyl bands, revealing oxidative modifications in the polymer structure (Rojas-Guerrero et al. 2023). The fading of bands between 2844 and 3050 cm^{-1} indicates structural changes in PS, confirming its progressive degradation during treatment (see Fig. 19.3).

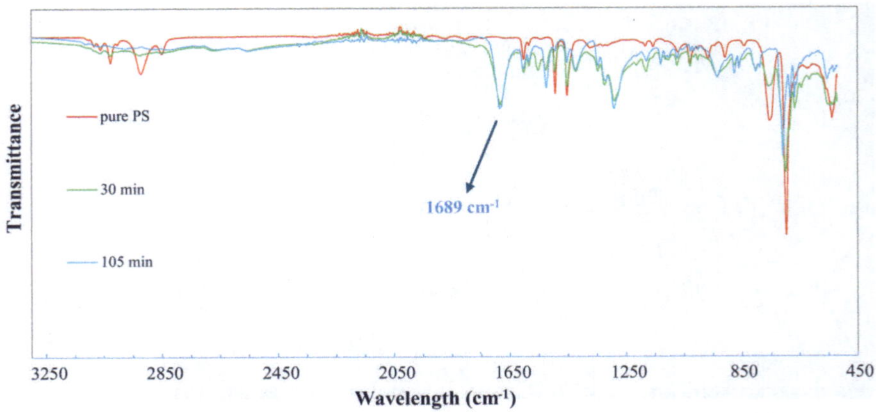

Fig. 19.3 FTIR spectra of polystyrene before and after degradation treatment

Figure 19.4 shows the degradation of PS particles produced by the microwave-assisted hydrogen peroxide treatment. Rapid degradation occurred within 105 min, resulting in an 84% mass loss. Particle degradation can be attributed to the synergistic effect of hydrothermal hydrolysis (high temperatures help to stretch and break the carbon chains), the proton-rich environment, and the production of hydroxyl radicals (Yang et al. 2024). Preliminary kinetic analysis indicates that PS particle degradation follows a pseudo-first-order kinetic model with a rate constant of (0.015 $\pm$ 0.001) min^{-1}. This pseudo-rate law has also been observed in the degradation of other microplastics using different advanced oxidation processes (Lin et al. 2024; Hu et al. 2022; Easton et al. 2023). The pseudo-first-order kinetic behavior is typical for microplastic chain scission caused by the attack of free radicals, which break the long chains of the plastic polymers and reduce particle size.

Total Organic Carbon (TOC) was measured in the liquid fraction from PS microplastic treatment. After 105 min of microwave-assisted degradation, TOC reached 15,574 $\pm$ 724 mg/L C, compared to 364 $\pm$ 15 mg/L C at 0 min. This sharp increase suggests the release of soluble organic compounds, potentially including toxic byproducts. While the elevated TOC confirms effective polymer breakdown, it also highlights the need to assess the chemical nature and environmental impact of the resulting liquid phase. These findings support the potential of this technique for sustainable plastic waste management, if toxicity risks are addressed.

Conclusions

The study demonstrates the effective degradation of PS microplastics using a microwave-assisted oxidative process with hydrogen peroxide and nitric acid, achieving an 84% mass loss within 105 min. Morphological and chemical analyses

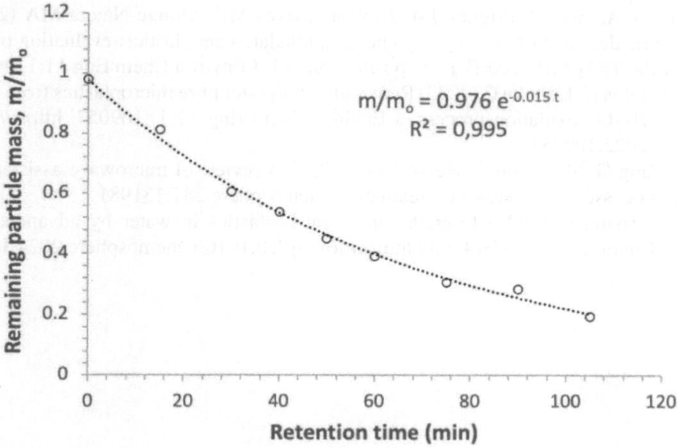

Fig. 19.4 Degradation of PS particles produced by microwave-assisted hydrogen peroxide treatment

reveal significant transformations in PS particles, including a change in color and structure, supported by mass loss measurements and FTIR spectroscopy showing carbonyl bands indicative of oxidative degradation. The results suggest a pseudo-first-order kinetic model for degradation, highlighting the potential of this innovative approach as a sustainable strategy for mitigating plastic pollution. Further investigation is required to elucidate the specific kinetics involved and validate the mechanistic hypotheses as well determine the chemical composition of the subproducts of the PS degradation.

Acknowledgements The authors gratefully acknowledge the University of Costa Rica (UCR) and the Research Center for Science and Engineering of Materials (CICIMA) for their support infrastructure that made this research possible.

References

Hu K, Zhou P, Yang Y, Hall T, Nie G, Yao Y, Duan X, Wang S (2022) Degradation of Microplastics by a thermal fenton reaction

Lin J, Hu K, Wang Y et al (2024) Tandem microplastic degradation and hydrogen production by hierarchical carbon nitride-supported single-atom iron catalysts. Nat Commun 15:8769. https://doi.org/10.1038/s41467-024-53055-1

Mdluli NS, Knottenbelt CD, Nomngongo PN, Mketo N (2024) Microwave-assisted hydrogen peroxide digestion followed by ICP-OES for determination of metals in selected fuel oils. Sci Rep 14(1):2362

Schrank I, Möller JN, Imhof HK, Hauenstein O, Zielke F, Agarwal S, Laforsch C (2022) Microplastic sample purification methods-assessing detrimental effects of purification procedures on specific plastic types. Sci Total Environ 833:154824

Rojas-Guerrero CA, Arce-Rodríguez JM, Alpízar-Chaves ME, Monge-Nájera MA (2023) Solar photocatalytic degradation of polyethylene terephthalate nanoplastics: evaluation of the applicability of the TiO_2/MIL-100(Fe) composite material. J Environ Chem Eng 11:110415
Thomas E, Vasileios K, Efthalia C (2023) Removal of polyester fibre microplastics from wastewater using a UV/H_2O_2 oxidation process. J Environ Chem Eng 11(1):109057. https://doi.org/10.1016/j.jece.2022.109057
Xia H, Li C, Yang G, Shi Z, Jin C, He W, Li G (2022) A review of microwave-assisted advanced oxidation processes for wastewater treatment. Chemosphere 287:131981
Yang Z, Li Y, Zhang G (2024) Degradation of microplastics in water by advanced oxidation processes. Chemosphere 357:141939.https://doi.org/10.1016/j.chemosphere.2024.141939

Chapter 20
Biodegradable but Persistent? Insights into the Environmental Implications of Microplastics from Biodegradable Plastics

Gabriela Kalčíková, Ula Putar, Lukas Brunnbauer, Andreas Limbeck, Jiří Kučerík, Petra Procházková, and Anita Jemec Kokalj

Abstract Biodegradable plastics are widely promoted as sustainable alternatives, yet their environmental behavior remains insufficiently understood. Our research shows that their degradation is low to moderate, strongly dependent on environmental conditions. Ecotoxicological tests revealed low acute but increased chronic effects in aquatic systems. In terrestrial environments, biodegradable microplastics showed similar effects to conventional plastics, including impacts on soil organisms and key soil properties. These findings highlight that "biodegradable" does not necessarily reflect actual degradation in the environment. Further testing under environmentally relevant conditions would help to improve our understanding and support informed decisions in material development and environmental policy.

G. Kalčíková (✉) · U. Putar
Faculty of Chemistry and Chemical Technology, Ljubljana, Slovenia
e-mail: gabriela.kalcikova@fkkt.uni-lj.si

G. Kalčíková
Faculty of Mechanical Engineering, Brno University of Technology, Brno, Czech Republic

L. Brunnbauer · A. Limbeck
TU Wien, Institute of Chemical Technologies and Analytics, Vienna, Austria

J. Kučerík
Department of Agrochemistry, Soil Science, Microbiology and Plant Nutrition, Faculty of AgriSciences, Mendel University in Brno, Brno, Czech Republic

P. Procházková
Central European Institute of Technology, Brno University of Technology, Brno, Czech Republic

A. Jemec Kokalj
Biotechnical Faculty, University of Ljubljana, Ljubljana, Slovenia

G. Gentile et al. (eds.), *Proceedings of the 4th International Conference on Microplastic Pollution in the Mediterranean Sea*, Springer Water,
https://doi.org/10.1007/978-3-032-23847-4_20

Introduction

Biodegradable plastics are increasingly promoted as sustainable alternatives to conventional plastics, which, due to their persistence and widespread use, have caused extensive pollution in both aquatic and terrestrial ecosystems (Napper and Thompson 2023). However, the actual behavior of biodegradable plastics in natural systems remains poorly understood. Recent studies have shown that degradation of biodegradable plastics is influenced by multiple factors, including polymer composition, additives, environmental conditions, and microbial activity (Afshar et al. 2024). Because biodegradable plastics are designed to meet specific laboratory or certification requirements for degradation, their degradation in real environmental conditions is not guaranteed, and in some cases, they may also fragment into microplastics (Convertino et al. 2024). Moreover, microplastics can then be transferred between environments, for example from soils to waterways, raising concerns about their persistence and ecological impact in aquatic and terrestrial systems (Crossman et al. 2020).

Understanding the environmental fate and potential ecotoxicological effects of microplastics derived from biodegradable plastics is crucial. Over the past years, our research has addressed these questions, investigating both terrestrial and aquatic ecosystems to provide insights into the sustainability and risks associated with these materials.

Materials and Methods

We investigated microplastics from two types of materials: polybutylene adipate terephthalate (PBAT) and poly(3-hydroxybutyrate) (PHB). PBAT microplastics (hereafter referred to as PBAT) were prepared by milling agricultural mulching films, while PHB microplastics (hereafter referred to as PHB) were purchased. The median size (absolute deviation) of PBAT microplastics was 147 ± 44 μm (Jemec Kokalj et al. 2024), while PHB fractions separated by sieving between 63 and 125 μm were used (Procházková et al. 2023). Detailed protocols for the preparation and characterization of these microplastics, including size distribution, morphology, and chemical properties, have been described in our previous studies (Jemec Kokalj et al. 2024; Procházková et al. 2023). Biodegradability testing was performed using a standard procedure for measuring oxygen consumption, according to ISO 14851:2019 to determination of the ultimate aerobic biodegradability of PBAT in an aqueous medium by measuring oxygen demand, with OxiTop respirometers. Prior biodegradability testing PBAT was pre-exposed to UV radiation (details are given in Putar et al. 2025) to simulate environmental weathering, as its properties can be significantly altered during use in agricultural fields.

The aquatic ecotoxicity of PHB microplastics was assessed using *Daphnia magna* in acute, chronic, and multi-generation experiments. Acute tests were conducted according to OECD guideline No. 202 up to 96 h. Chronic and multi-generation exposures followed OECD guidelines No. 211, assessing survival, growth, reproduction, and offspring production over 21 days and across four consecutive generations (Procházková et al. 2024). The test with *Lemna minor* was performed according to OECD guidelines No. 221 and specific growth rate, root length, and content of photosynthetic pigments were determined (Procházková et al. 2023). Effects on terrestrial organisms were assessed using woodlice (*Porcellio scaber*), which were exposed to PBAT in test soil along with a sterilized hazel leaf as food. Survival, immune parameters, and metabolic activity were monitored, and effects on soil properties, including pH and water holding capacity (WHC), were also evaluated (Jemec Kokalj et al. 2024).

Results and Discussion

Biodegradation in the Aquatic Environment

PHB is a bio-based polymer well recognized for its efficient biodegradation and is commonly employed as a reference compound in biodegradability tests (e.g. in ISO 14851:2019); consequently, its degradation in aquatic environments is not regarded as problematic. On the other hand, PBAT is a fossil-based polymer, and its biodegradability and physicochemical changes have not been systematically assessed. The maximal biodegradation of pristine PBAT reached 15%, while UV pre-treatment markedly enhanced the biodegradation up to 57%. Morphological analysis revealed distinct changes: pristine PBAT displayed a relatively rough surface, whereas UV-aged PBAT became smoother and after the biodegradation test additional formation of small surface holes was confirmed. Elemental analysis indicated leaching of additives, as evidenced by the loss of titanium in UV-aged PBAT after biodegradation (Putar et al. 2025). These findings suggest that although some structural and compositional changes occur, PBAT cannot be considered fully biodegradable under the tested conditions. While materials such as mulching foils, are designed to degrade in soil environments, their persistence in aquatic systems is likely driven by lower microbial activity and limited nutrient availability. Moreover, the presence of stabilizing additives, although beneficial for mechanical performance in agriculture, appears to reduce the overall degradation potential. Nevertheless, our results confirm that UV exposure plays a critical role in enhancing the biodegradability of PBAT microplastics.

Effects on Aquatic and Terrestrial Organisms

The ecotoxicity of PHB and PBAT was assessed using *Lemna minor* at a concentration of 100 mg/L, which is considered the standard limit for ecotoxicological studies. PBAT showed no effects on *Lemna minor* across any of the evaluated endpoints (Putar et al. 2025). In contrast, PHB caused a significant effect on root growth. Specifically, root elongation was observed, with roots being approximately 1.70 times longer in the PHB treatment compared to the control. This effect was attributed to the intensive biodegradation of PHB, which enhanced microbial activity and led to increased utilization of available nutrients essential for plant growth (Procházková et al. 2023). As a result, the plants could compensate for nutrient limitation by producing longer roots (Hermans et al. 2006). Further testing was conducted with PHB to evaluate its effects on *Daphnia magna*, focusing on acute, chronic, and multigenerational endpoints, including inhibition, mortality, reproductive activity, and growth. No acute effects were observed, even after 96 h of exposure at concentrations up to 100 mg/L (Procházková et al. 2024). For chronic and multigenerational experiments, lower concentrations (1.56–25 mg/L) were applied due to the high mortality observed in preliminary tests at higher concentrations. A slight reduction in reproductive activity (12.2%) was detected only at 25 mg/L of PHB. Additionally, a slight delay in the average day of the first brood release (up to 14.2%) was observed. The growth rate of maternal organisms showed a minor stimulation at the lowest concentration tested (2.1%), followed by a slight inhibition of growth at higher concentrations (up to 8.6%). In the multigenerational experiment, conducted at 1.56 mg/L, growth stimulation was observed in the first generation (F0) of maternal organisms (4.0%). However, in subsequent generations, a progressive inhibition of reproductive activity was recorded, reaching 19.2% in the F3 generation (Procházková et al. 2023). These findings demonstrate that while acute tests did not reveal measurable effects, chronic and multigenerational exposures resulted in significant impacts, comparable or higher to those reported for microplastics derived from conventional polymers (Savva et al. 2023).

Since PBAT is used as mulching film, it can be directly fragmented into soil; therefore, we tested its effects on the terrestrial isopod *Porcellio scaber*. The concentrations in soil were 0.005, 0.05, 0.1, 0.5, 1, and 5% (w/w dry soil). No changes in survival were observed after two weeks of exposure. However, metabolic activity, measured as electron transport system (ETS) activity, increased at 1% and 5% PBAT and changes in ratios between different immune cells were observed already at 1% PBAT. At the end of the experiment, soil pH and WHC were measured. Surprisingly, 5% of PBAT significantly affected both parameters: pH decreased from 5.86 to 5.60, while WHC increased from 42 to 52% (Jemec Kokalj et al. 2024). These changes may have long-term implications, as soils will continue to accumulate microplastics over time, and alterations in WHC could increase the need for irrigation, which is particularly concerning in the context of climate change.

Conclusions

These results indicate that the term "biodegradable" does neither reflect the actual degradation behavior of plastics in natural environments, nor their safety for organisms in aquatic and terrestrial systems. Further testing under environmentally relevant conditions is needed to improve to understanding of their environmental impacts and to support informed decisions in material development and environmental policy.

Acknowledgements Slovenian Research and Innovation Agency (ARIS) (J1-50014 "MicroBIOplast", research programs P1-0184 and P2-0191).

References

Afshar SV, Boldrin A, Astrup TF, Daugaard AE, Hartmann NB (2024) Degradation of biodegradable plastics in waste management systems and the open environment: a critical review. J Clean Prod 434:140000. https://doi.org/10.1016/j.jclepro.2023.140000

Convertino F, Carroccio SC, Cocca MC, Dattilo S, Dell'Acqua AC, Gargiulo L, Nizzetto L, Riccobene PM, Schettini VG, Zannini D, Cerruti P (2024) The fate of post-use biodegradable PBAT-based mulch films buried in agricultural soil. Sci Total Environ 948:174697. https://doi.org/10.1016/j.scitotenv.2024.174697

Crossman J, Hurley RR, Futter M, Nizzetto L (2020) Transfer and transport of microplastics from biosolids to agricultural soils and the wider environment. Sci Total Environ 724:138334. https://doi.org/10.1016/j.scitotenv.2020.138334

Hermans C, Hammond JP, White PJ, Verbruggen N (2006) How do plants respond to nutrient shortage by biomass allocation? Trends Plant Sci 11:610–617. https://doi.org/10.1016/J.TPLANTS.2006.10.007

Jemec Kokalj A, Dolar A, Nagode A, Drobne D, Kuljanin A, Kalčíková G (2024) Response of terrestrial crustacean Porcellio scaber and mealworm Tenebrio molitor to non-degradable and biodegradable fossil-based mulching film microplastics. Sci Total Environ 951. https://doi.org/10.1016/j.scitotenv.2024.175379

Napper IE, Thompson RC (2023) Plastics and the environment. Annu Rev Environ Resour 48:55–79. https://doi.org/10.1146/annurev-environ-112522-072642

Procházková P, Mácová S, Aydın S, Zlámalová Gargošová H, Kalčíková G, Kučerík J (2023) Effects of biodegradable P3HB on the specific growth rate, root length and chlorophyll content of duckweed, Lemna minor. Heliyon 9:e23128. https://doi.org/10.1016/j.heliyon.2023.e23128

Procházková P, Kalčíková G, Maršálková E, Brtnický M, Zlámalová Gargošová H, Kučerík J (2024) Assessing the ecological consequences of biodegradable plastics: acute, chronic and multigenerational impacts of poly-3-hydroxybutyrate microplastics on freshwater invertebrate Daphnia magna. Heliyon 10:e36302. https://doi.org/10.1016/j.heliyon.2024.e36302

Putar U, Fazlić A, Brunnbauer L, Novak J, Jemec Kokalj A, Imperl J, Kučerík J, Procházková P, Federici S, Hurley R, Sever Škapin A, Modlitbová P, Pořízka P, Kaiser J, Limbeck A, Kalčíková G (2025) Investigating aquatic biodegradation and changes in the properties of pristine and UV-irradiated microplastics from conventional and biodegradable agricultural plastics. Env Pollut 376: 126408. https://doi.org/10.1016/j.envpol.2025.126408

Savva K, Farrè M, Barata C (2023) Sublethal effects of bio-plastic microparticles and their components on the behaviour of Daphnia magna. Environ Res 236:116775. https://doi.org/10.1016/j.envres.2023.116775

Chapter 21
Right of Consumers to Information: What Should We Know About Plastic?

Esther Kentin and Xenia van Tuinen

Introduction

This contribution examines the current state of consumer information on microplastics and plastic composition in products in the European Union (EU). Despite the EU's commitment to a high level of consumer protection, there remain substantial gaps in transparency regarding the materials and additives used in everyday plastic products. The current legal framework on consumer information as well as EU plastic regulation does not require producers to disclose detailed, plastic-related information. Promisingly, the forthcoming Ecodesign for Sustainable Products Regulation (ESPR) and the Digital Product Passport (DPP) offer opportunities to enhance transparency, provided they impose specific, enforceable disclosure obligations for plastic-containing product groups.

The Right to Information in European Consumer Law

A high level of consumer protection is the underlying premise of the European social market economy, one of the main objectives of the European Union included in Article 38 Charter of Fundamental Rights. A fundament of high protection is transparency and accessibility of information for the consumer, as is highlighted in the EU Consumers Rights Directive and case law (Cassis de Dijon). Article 5 of this directive requires the seller to provide the consumer with information about the main characteristics of goods in a clear and comprehensible manner. The right to information is seen essential as both providing protection of consumers' interests

E. Kentin (✉) · X. van Tuinen
Leiden Law School, Leiden University, Leiden, Netherlands
e-mail: e.kentin@law.leidenuniv.nl

G. Gentile et al. (eds.), *Proceedings of the 4th International Conference on Microplastic Pollution in the Mediterranean Sea*, Springer Water,
https://doi.org/10.1007/978-3-032-23847-4_21

as to trigger the right of redress. Case law confirms that the essential guarantee of fairness of commercial transactions could be ensured by requiring the display of suitable information on the packaging of products based on the idea that the average consumer is a 'reasonably circumspect and well-informed consumer' (Gut Springenheide and Sektkellerei Kessler). Furthermore, the Unfair Commercial Practices Directive requires that consumers not be deceived by omission or concealment of information, deceiving the average consumer and leading to a purchase that would not have taken place otherwise (Nehf 2016).

The question is whether information about plastic would be considered by the consumer as a main characteristic of the product and whether this would affect the choice of the consumer. Studies suggest that the availability of specialist information is a factor for making a more sustainable choice (McKinsey and NielsenIQ 2023). Consumers are increasingly aware of plastic pollution and plastic waste, as part of global environmental concerns (Bastos de Sousa 2023), although it seems that general knowledge about plastic pollution does not automatically lead to changing habits (SEA circular project and Food Industry Asia 2020). Consumer organisations have addressed plastic packaging and products in their campaigns as part of the issue of sustainable consumption and some studies suggest that consumers are willing to pay for more sustainable packaging and that they rely on information that is immediately available (Herrmann 2022; Young 2010; Ran 2022).

Do Product Standards, Requirements, and Labelling Inform the Consumer?

So information about the composition of a product and its packaging can be essential for consumers in making choices. The subsequent question is whether consumers are sufficiently informed about 'the main characteristics of goods in a clear and comprehensible manner' and about 'their nature and composition' as required by EU regulation. The European Union has a long history of establishing standards and requirements for different product groups, primarily to ensure safety, health, and environmental protection, but the transparency and accessibility of detailed information for the average consumer is not always guaranteed and depends on the regulation of the specific product. We will now discuss several product policies for which requirements exist regarding the presence of plastics with a focus on the availability of information for consumers.

The Regulation concerning the Registration, Evaluation, Authorisation and Restriction of Chemicals (REACH) regulates chemicals in order to protect human health and is primarily targeted to manufacturers, importers, and downstream users. An exception to this approach is Article 33 which provides consumers with a right to request information on substances of very high concern (SVHCs) in consumer products. However, not many consumers are aware of this right and although there are some SVHCs that may qualify as plastic, most plastics will not be considered as such. Under REACH, intentionally microplastics will be gradually phased out, in many consumer products, such as cosmetics and detergents, with final dates in 2035, though this measure has not been communicated to consumers directly. Only after 2031 labelling is required for products still containing microplastics.

In the same vein is the Food Contact Materials regulation of 2004, amended in 2011 regarding plastics. While regulating plastic materials to protect human health, the regulation does not foresee to inform the consumer, as only requirements regarding information to producers and downstream business operators are included.

The Single Use Plastic directive of 2019 introduced measures for ten single use plastic products, of which the plastic straws, bottles and cups were most prominent. While banning some products, the directive includes obligations to raise awareness of plastic content in other products and requires marking to inform consumers accordingly, for example, on coffee cups, cigarettes, and wet wipes, where the presence of plastic is less apparent.

The new Packaging and Packaging Waste Regulation (PPWR) of 2025 focuses on preventing packaging waste and promoting re-use and recycling. Regarding information on plastic in packaging, the PPWR requires, by 2028, that packaging is marked with a harmonised label containing information on its material composition so consumers can appropriately dispose of. The design should be recognisable for 'all citizens, irrespective of their circumstances, such as age and language knowledge'.

The Cosmetics Regulation of 2009 introduced the obligation to list the ingredients of the product on the product or packaging available for consumers. Consumers are therefore able to find out whether the product contains plastic. Whether this is true for the 'average consumer' may be questioned as ingredients are stated according to the International Nomenclature of Cosmetic Ingredients (INCI), requiring specialist knowledge to identify plastics.

Due to the European Textile regulation (2011/2017), which regulates fibre names, and the labelling and marking of fibre composition of textile products, consumers can identify synthetic fabrics, made out of polymers. It is estimated that around 60% of fabrics consists of synthetic fibres and that this amount is continuing to grow. What a label does not reveal is whether and how the fabric has been treated. Colouring, additives, coatings or any other substances used on the fabric, are not required to be indicated. The 2022 EU Strategy for Sustainable and Circular Textiles addresses the shedding of microplastics from synthetic clothing and foresees in 'creating conditions and incentives to boost the competitiveness, sustainability and resilience of the EU textiles sector'.

The new Ecodesign for Sustainable Products Regulation (2024/1781) (ESPR) is applicable to all physical products and extends the range of ecodesign requirements beyond energy efficiency. New requirements can be set on durability, reusability, reparability, circularity, carbon and environmental footprints, generation of waste and other sustainability benchmarks for specific product groups. With access to reliable and clear data, consumers and industry will be able to make more sustainable choices. A further, important component is the introduction of the Digital Product Passport (DPP), which will include information on substances of concern and on sustainability requirements. The DPP will be publicly available for all stakeholders and, once realised in an ambitious manner, will significantly improve access to information for consumers in the EU. However, it remains to be seen whether detailed information about plastics, including additives, will be included in the DPP. As things stand today, the plan of the Commission includes only 'substances of concern' and most plastics are not considered as such.

Conclusions

The present regulation regarding plastics in products is fragmented, including the information available for consumers. While microplastics in some products are being regulated in by REACH, this does not mean that the presence of microplastics is communicated to consumers. Protection of health of consumers, safety and environmental protection does not coincide with the availability of information for consumers. We conclude that there is a significant gap between the European Union's goal of high consumer protection and the limited information consumers actually receive regarding plastic products. Although EU consumer law mandates that producers and traders must communicate essential product characteristics, information concerning the composition and additives of plastic products is generally not available or difficult to access. The introduction of the Ecodesign for Sustainable Products Regulation and its Digital Product Passport offers a promising route to mandate information on plastic content, but its effectiveness depends on the subsequent adoption of concrete, product-specific requirements. Ultimately, further disclosure from producers on plastic composition is required to enable consumers to make genuinely informed and sustainable purchasing choices—and to fulfil the promise of a high level of consumer protection in the Charter of Fundamental Rights of the EU.

References

EU Regulations and Directives

Regulation 2025/40/EU on Packaging and Packaging Waste (2025) OJ L 2025/40

Regulation 2024/1781/EU on Ecodesign Requirements (2024) OJ L 2024/1781

Directive 2019/904 on Single-Use Plastics (2019) OJ L 155/1

Directive 2011/83/EU on Consumer Rights (2011) OJ L 304/64

Regulation 1007/2011 on Textile Fibre Names and Related Labelling and Marking of the Fire Composition of Textile Products (2011) OJ L 272/1

Regulation 10/2011 on Food Contact Materials (2011) OJ L 12

Regulation 2009/1223/EC on Cosmetic Products (2009) OJ L 342/59

Regulation 2006/1907/EC on Registration, Evaluation, Authorisation, Restriction of Chemicals (2006) OJ L 396/1

Directive 2005/29/EC on Unfair Commercial Practices (2005) OJ L 149/22

Case Law

Case C-303/97 Sektkellerei Kessler [1999] ECR I-513
Case C-210/96 Gut Springenheide GmbH and Rudolf Tusky v. Oberkreisdirektor des Kreises Steinfurt—Amt für Lebensmittelüberwachung [1998] ECR I-4657
Case 120/78 Rewe-Zentral AG v Bundesmonopolverwaltung für Branntwein (Cassis de Dijon) [1979] ECR 649

Literature

Bastos de Sousa FD (2023) Consumer awareness of plastic: an overview of different research areas. Circ Econ Sustain 3:2083–2107
Herrmann C et al (2022) Consumers' sustainability-related perception of and willingness-to-pay for food packaging alternatives. Resour Conserv Recycl 181. https://doi.org/10.1016/s106219
McKinsey and NielsenIQ (2023) Consumers care about sustainability—and back it up with their wallets, McKinsey Global Publishing
Nehf JP (2016) Misleading and unfair advertising. In: Howells G (eds) Handbook of research on international consumer law, PP 107–129
Ran Y (2022) Information as an enabler of sustainable food choices: a behavioural approach to understanding consumer decision-making. Sustain Prod Consum 31(1):642–656
SEA circular project and Food Industry Asia (2020) Perceptions on plastic waste, insights, interventions and incentivces to action from businesses and consumers in South-East Asia
Young W (2010) Sustainable consumption: green consumer behaviour when purchasing goods. Sustain Dev 18(1):20–31

Chapter 22
Exploring the potential of One-Pot Microwave-Assisted Pretreatments Coupled With PY-GC-MS for the Quantification of MPs and Associated Contaminants

Jacopo La Nasa, Greta Biale, Adrián Jaén-Gil, Lucia Pittura, Marco Mattonai, Diego Carnaroglio, Luca Rota, Stefania Gorbi, Francesco Regoli, Francesca Modugno, and Alessio Gomiero

Introduction

Microplastic (MP) pollution is now recognized as a global environmental threat and one of the most urgent challenges for ecosystems and human health. MPs persist in the environment and act as carriers of chemical species (Wang 2020) including additives, persistent organic pollutants, and polymer degradation products. Despite numerous studies, the interactions between MPs, associated contaminants, organisms and ecosystems remain insufficiently understood (Barcelò 2023; Thompson

---

J. La Nasa · G. Biale · M. Mattonai · F. Modugno
Department of Chemistry and Industrial Chemistry, University of Pisa, Pisa, Italy

J. La Nasa (✉) · G. Biale · F. Modugno
Center for Instrument Sharing of the University of Pisa, Pisa, Italy
e-mail: jacopo.lanasa@unipi.it

A. Jaén-Gil · A. Gomiero
Department of Climate and Environment, Norwegian Research Centre, Stavanger, Norway

L. Pittura · S. Gorbi · F. Regoli
Department of Life and Environmental Sciences, Polytechnic University of Marche, Ancona, Italy

NBFC, National Biodiversity Future Center, Palermo, Italy

D. Carnaroglio · L. Rota
Milestone Srl, Sorisole, Italy

G. Gentile et al. (eds.), *Proceedings of the 4th International Conference on Microplastic Pollution in the Mediterranean Sea*, Springer Water,
https://doi.org/10.1007/978-3-032-23847-4_22

2024). A variety of analytical techniques have been developed to study MPs and co-occurring pollutants. Among these, pyrolysis-based methods have proven particularly effective, as they allow mass-based quantification of different polymers together with the identification of associated organic contaminants (La Nasa 2020; Akoueson 2021). However, the analysis often requires labor-intensive pretreatments that vary depending on the complexity of the environmental or biological matrix (Lykkemark 2024). To address the limitations of conventional multi-step sample preparation, we evaluated microwave-assisted extraction and digestion as integrated and efficient alternatives. The approach combines microwave-assisted extraction with thermal desorption and pyrolysis–gas chromatography–mass spectrometry (Py-GC–MS), enabling the characterization and quantification of several classes of pollutants, including phthalate plasticizers, polycyclic aromatic hydrocarbons, polychlorinated biphenyls, and contaminants of emerging concern. In parallel, the same samples were subjected to microwave-assisted digestion, which was optimized to reduce polymer degradation while allowing reliable quantification of microplastics. This combined workflow provides a rapid and comprehensive overview of both polymeric and non-polymeric contaminants within complex biological matrices. Even though current methods enable the characterization of microplastics (MPs) and associated contaminants, they generally do not allow the analysis of both classes of compounds from the same sample. To overcome this limitation, a one-pot microwave-assisted pretreatment was optimized to provide a comprehensive overview of MPs and co-occurring contaminants within a single workflow. Mussels exposed to polypropylene (PP), polyethylene (PE), and nylon-6 (N6) were used as a model matrix, chosen for their high protein and lipid content as well as their importance as bioindicators of marine pollution. This rapid and integrated method reduces sample preparation time, improves efficiency, and opens the way for applications to other biological matrices, including human samples. Such an approach is essential for assessing MP exposure and co-contaminant burdens and has direct implications for environmental risk assessment and public health.

Materials and methods

Microwave assisted extraction and digestion

All digestion and extraction procedures were performed using an Ethos X Advanced Microwave Extraction System equipped with a SK-15high-pressure rotor (Milestone Srl, Italy). The conditions used for the solvent-extractable plastics are reported in Biale et al. (2024), while the conditions for the parallel digestion experiments are reported in La Nasa et al. (2021).

Py-GC-MS analysis

The analyses were performed using a multi-shot pyrolyzer EGA/PY-3030D (Frontier Lab, Japan) coupled with an 8890 gas chromatograph, combined with a 5977B mass selective single quadrupole mass spectrometer detector (Agilent Technologies, US). The complete conditions used for the studies are reported in Biale et al. (2024) and La Nasa et al. (2021). The analysis for the one-pot sample pretreatment was performed using a multi-shot pyrolyzer EGA/PY-3030D (Frontier Lab) coupled with a Shimadzu GCMS-QP2020 NX instrument (Shimadzu, Japan).

Results and discussion

Microplastic (MP) pollution poses major environmental concern, not only because MPs persist in ecosystems but also because they carry and release a variety of chemical contaminants such as additives, degradation products, and persistent organic pollutants. Robust analytical workflows are therefore required to quantify MPs together with associated chemicals in different environmental and biological matrices. Traditional procedures are often based on multi-step digestion and separation, which are labor-intensive and may lead to polymer losses or degradation.

Recent work has demonstrated microwave-assisted pretreatments as a faster and more efficient alternative. Microwave energy accelerates both extraction and digestion, reduces solvent consumption, and improves reproducibility. In one study, microwave-assisted solvent extraction was coupled with double-shot pyrolysis–gas chromatography–mass spectrometry (Py-GC–MS) to analyze MPs and additives in beach sand (Fig. 22.1). This method allowed volatile compounds such as phthalates to be released at 350 °C and polymers to be characterized at 600 °C. Recoveries were above 96% for phthalates and 97% for polystyrene, with detection limits in the nanogram range. Application to sand samples from the Tuscan Coast revealed heterogeneous distributions of MPs and additives, showing the influence of leaching and degradation processes. Insoluble polymers like polyethylene and polypropylene could not be directly quantified, but their oxidation products were detected, highlighting the sensitivity of the approach. More recently, the method was adapted to biological matrices such as mussels, which are rich in proteins and lipids and widely used as sentinel organisms for marine pollution (Fig. 22.2). The approach combined two parallel microwave steps: solvent extraction of organic pollutants and acid digestion for MPs, followed by Py-GC–MS. This workflow enabled the quantification of more than 40 contaminants, including phthalates, polycyclic aromatic hydrocarbons, polychlorinated biphenyls, and contaminants of emerging concern, along with MPs from a calibration set of 11 polymers. Recoveries were generally above 87%, and detection limits remained in the nanogram range. Hydrochloric acid digestion gave the most reliable results, while some polymers such as PET and nylon showed poor or overestimated recoveries.

Together, these studies show the evolution of microwave-assisted sample preparation from environmental solids to complex biological tissues. The integration of microwave pretreatments with analytical pyrolysis provides a rapid and comprehensive tool for the analysis of MPs and co-occurring contaminants. This approach

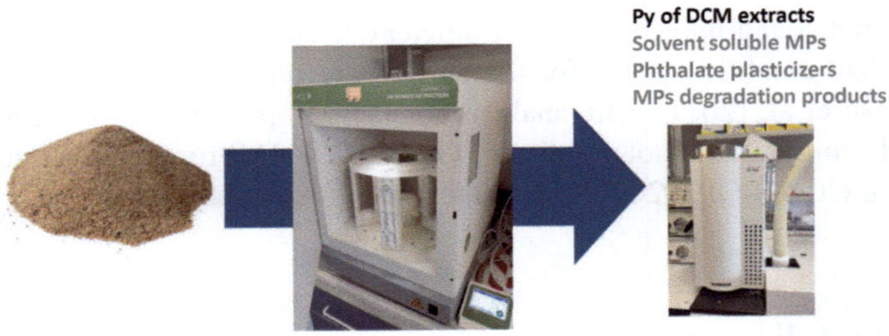

Fig. 22.1 Summary of the MW-assisted procedure for the study of MPs and pollutants in sand

Fig. 22.2 Summary of the MW-assisted procedure for the study of MPs and pollutants in mussels

reduces preparation time, minimizes matrix interferences, and has potential applications in environmental monitoring, seafood safety, and future human exposure assessment. The study further optimized a one-pot protocol for the extraction of phthalates and the digestion of microplastics in seafood samples (Fig. 22.3). Cyclohexane was selected as the extraction solvent for its chemical stability and ease of removal, yielding recoveries above 95% for most phthalates, However, lipid-rich extracts can reduce GC-MS sensitivity, so an additional purification step with Enhanced Matrix Removal–Lipid cartridges was tested. While effective at reducing lipids, this step caused significant losses of high-molecular-weight phthalates, limiting its use to preliminary screenings. For microplastic isolation, digestion protocols using HCl/H_2O_2 and HNO_3/H_2O_2 were compared under microwave conditions. HNO_3-based digestion provided the cleanest results, with 120 °C identified as the optimal condition to ensure polymer preservation while minimizing matrix residues.

The optimized protocol therefore balances phthalate recovery, lipid removal, and microplastic stability, enabling reliable analysis of seafood samples.

Conclusions

This work demonstrates the feasibility and advantages of microwave-assisted protocols for the integrated analysis of microplastics and associated contaminants in complex biological matrices. Microwave-assisted extraction proved effective for phthalates and other organic pollutants, whereas microwave-assisted digestion

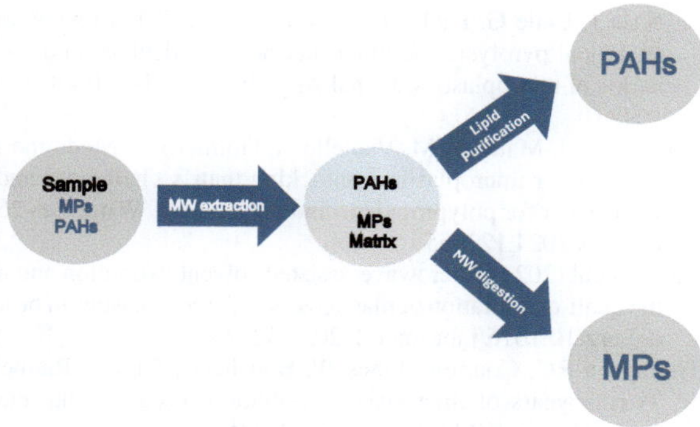

Fig. 22.3 Summary of the one-pot MW-assisted protocol

enabled the isolation and quantification of different polymer types, and their integration into a one-pot workflow allowed complementary measurements to be obtained from a single sample. Although each approach presents specific advantages and limitations, together they illustrate how microwave technologies can streamline sample pretreatment, reduce analysis time, and improve reproducibility. Their flexibility supports applications across diverse environmental and biological matrices, contributing to more comprehensive assessments of microplastic pollution and co-occurring chemical exposures.

Acknowledgement This work was carried thanks to the financial support of NAMC (North Atlantic Microplastic Centre), the funding from the European Union-Next-GenerationEU-National Recovery and Resilience Plan (NRRP)—MISSION 4 COMPONENT 2, INVESTIMENT N. 1.1, CALL PRIN 2022 D.D. 104 02-02-2022—DIORAMA (A deep dive into the study of microplastics in aqueous matrices), and the National Recovery and Resilience Plan (NRRP), Mission 4 Component 2 Investment 1.4 of MUR by the European Union's NextGenerationEU (Project CN 00000033, Decree No. 1034 of June 17, 2022 adopted by the MUR, CUP I33C22001300007, Project title "National Biodiversity Future Center-NBFC"). We acknowledge the NORCE Mass Spectrometry Lab for instrument access and expert support.

References

Akoueson F, Chbib C, Monchy S, Paul-Pont I, Doyen P, Dehaut A, Duflos G (2021) Identification and quantification of plastic additives using pyrolysis-GC/MS: a review. Sci Total Environ 773:145073. https://doi.org/10.1016/j.scitotenv.2021.145073

Barceló D, Picó Y, Alfarhan AH (2023) Microplastics: detection in human samples, cell line studies, and health impacts. Environ Toxicol Pharmacol 101:104204. https://doi.org/10.1016/j.etap.2023.104204

Biale G et al (2024) Characterization and quantification of microplastics and organic pollutants in mussels by microwave-assisted sample preparation and analytical pyrolysis. Environ Sci Adv. https://doi.org/10.1039/D3VA00216K

La Nasa J, Biale G, Fabbri D, Modugno F (2020) A review on challenges and developments of analytical pyrolysis and other thermo-analytical techniques for the quali-quantitative determination of microplastics. J Anal Appl Pyrolysis 149:104841. https://doi.org/10.1016/j.jaap.2020.104841

Lykkemark J, Mattonai M, Vianello A, Gomiero A, Modugno F, Vollertsen J (2024) Py–GC–MS analysis for microplastics: unlocking matrix challenges and sample recovery when analyzing wastewater for polypropylene and polystyrene. Water Res 261:122055. https://doi.org/10.1016/j.watres.2024.122055

Nasa L et al (2021) Microwave-assisted solvent extraction and double-shot analytical pyrolysis for the quali-quantitation of plasticizers and microplastics in beach samples. J Hazard Mater. https://doi.org/10.1016/j.jhazmat.2020.123287

Thompson RC, Courtene-Jones W, Boucher J, Pahl S, Raubenheimer K, Koelmans AA (2024) Twenty years of microplastics pollution research-what have we learned? Science:eadl2746. https://doi.org/10.1126/science.adl2746

Wang T, Wang L, Chen Q, Kalogerakis N, Ji R, Ma Y (2020) Interactions between microplastics and organic pollutants: effects on toxicity, bioaccumulation, degradation, and transport. Sci Total Environ 748:142427. https://doi.org/10.1016/j.scitotenv.2020.142427

Chapter 23
Zebrafish to Study Microplastics and Trojan Horse Effect

Alessandra La Pietra, Teresa Mobilio, and Ida Ferrandino

Introduction

Microplastics are one of the most pressing environmental concerns in recent decades due to their widespread presence in ecosystems worldwide. Among the various types of microplastics, polystyrene is one of the most common polymers, often used for packaging materials and other everyday items. Polystyrene microplastics (PS-MPs) pose significant environmental risks, not only for their potential toxicity but also for their capacity to carry other chemical pollutants due to their hydrophobic surface (Siddiqui 2023). One such chemical is bisphenol A (BPA), a widely used industrial compound in the production of plastics and resin and known for its endocrine-disrupting properties (Rochester (2014). Microplastics and bisphenols are frequently found together at polluted sites and microplastics may be carriers of BPA, adsorbing it on their surfaces through hydrophobic interactions (Zhu 2022). In aquatic environments, fishes are directly exposed to microplastic particles bind with BPA and can ingest or absorb them.

This study analyzed polystyrene microplastics of 1 μm in size at 0.01, 0.1, 1, and 10 mgL^{-1}, and bisphenol A at 6.25, 12.5, 25, and 50 μM individually on zebrafish embryos, model organism widely used for toxicology research (Ferrandino 2024). Toxicity parameters such as survival, hatching, and heart rate were analyzed up to 72 h of exposure. Subsequently, a combination of PS-MPs (1 mgL^{-1}) and BPA (12.5 μM) was tested to evaluate how microplastics influence the action of BPA on zebrafish development, in particular on the heart development and in the expression of gene involved in it (*gata4*, *hand2* and *sox9b*) through quantitative real time PCR (qRT-PCR). In fact, it is known that BPA binds the estrogen receptors that are present also on the heart altering its physiological functioning (Moreman et al.

A. La Pietra (✉) · T. Mobilio · I. Ferrandino
Department of Biology, University of Naples Federico II, Naples, Italy
e-mail: alessandra.lapietra@unina.it

G. Gentile et al. (eds.), *Proceedings of the 4th International Conference on Microplastic Pollution in the Mediterranean Sea*, Springer Water,
https://doi.org/10.1007/978-3-032-23847-4_23

2017). The concentrations for co-exposure were selected based on the minimum toxic effect observed in the single exposure experiments to avoid excessive effects and to understand the interaction between the two pollutants.

Materials and Methods

Chemicals

Polystyrene microplastics (microParticles GmbH, Germany, density of 1.05 g/cm^3) solutions were prepared in E3 medium (5 mM NaCl, 0.17 mM KCl, 0.33 mM $CaCl_2 \cdot 2H_2O$, 0.33 mM $MgSO_4$) while bisphenol A (Sigma-Aldrich, France) was first dissolved in dimethyl sulfoxide (DMSO) and then diluted in E3 medium to reach concentration to use. For the co-exposure the solutions were prepared 24 h before the start of experiments to allow the bind between BPA and PS-MPs (Chen 2017).

Toxicity parameters

Ten zebrafish embryos for each group were exposed to the solutions that were renewed daily. One group was exposed only to E3 medium as negative control. In addition, DMSO solutions (0.0025, 0.005, 0.01, 0.02%) at the same concentration present in the BPA solutions were used as an additional negative control. Survival and hatch were followed up to 72 h of treatment. Heart rate was measured at 72 h of treatment counting it for 1 min using light microscope.

Quantitative real-time PCR

RNA was extracted from homogenized larvae and then retrotranscribed into cDNA. The qRT-PCR was set up to analyze expression of *gata4*, *hand2* and *sox9b* and *β-actin* was used as a housekeeping gene. The data were analyzed through the Ct value.

Statistical Analysis

All experiments were repeated in triplicate and statistical analysis was done using GraphPad Prism Software (version 8.02 for Windows, GraphPad Software, La Jolla, CA, USA). The lowest level of significance that could be accepted was set at $p < 0.05$.

Results and Discussion

Toxicity parameters during the single exposure to PS-MPs or BPA

The results showed that 1 μm PS-MPs at all concentrations tested did not affect survival or hatching (Fig. 23.1 a, b). Instead, the heart rate analysis revealed that PS-MPs at higher concentrations tested caused tachycardia in zebrafish larvae (Fig. 23.1c) as previously demonstrated by La Pietra et al. (2024). The presence

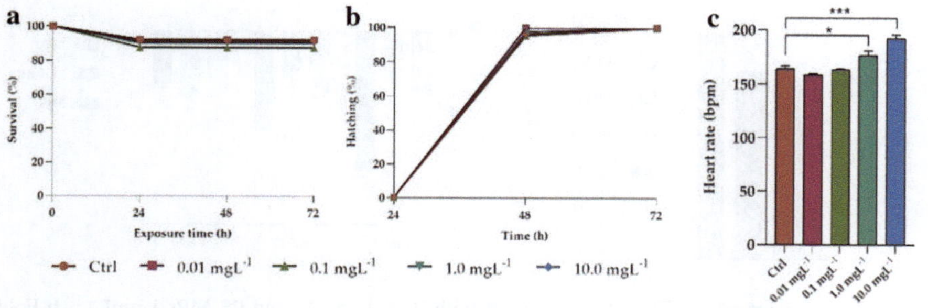

Fig. 23.1 Toxicity parameters during treatment with PS-MPs. **a** Survival; **b** Hatching rate; **c** Heart rate at 72 h of treatment with PS-MPs. Dunnett's test (*$p < 0.05$; *** $p < 0.001$)

of microplastics inside the larvae could cause hypoxic conditions with consequent alterations of blood flow and increase in heart rate (Lu etal. 2022).

From experiments with BPA, the results demonstrated that it reduced the survival and the hatching rate at higher concentrations tested (25 and 50 μM) (Fig. 23.2 a, b). These effects were previously observed by Moreman et al. (2017) and may be due to the binding of the BPA to thyroid hormones that are involved in development of zebrafish larvae (Vancamp 2019). In addition, BPA caused bradycardia at 12.5 and 25 μM (Fig. 23.2c) according to Qin et al. (2023) and this effect could be due to BPA binding to estrogen receptors in the heart, which could alter its normal functioning (Moreman 2017).

The DMSO did not cause toxic effects at any concentration.

Effects of PS-MPs on the toxic action of BPA at heart level

The data showed that during co-exposure, there was a slight decrease in the effects of two pollutants (Fig. 23.3a). This trend was partly confirmed by gene expression analysis. In fact, *hand2* was less down-regulated during co-exposure, while *gata4*, although not significant, showed a tendency to improve during co-exposure

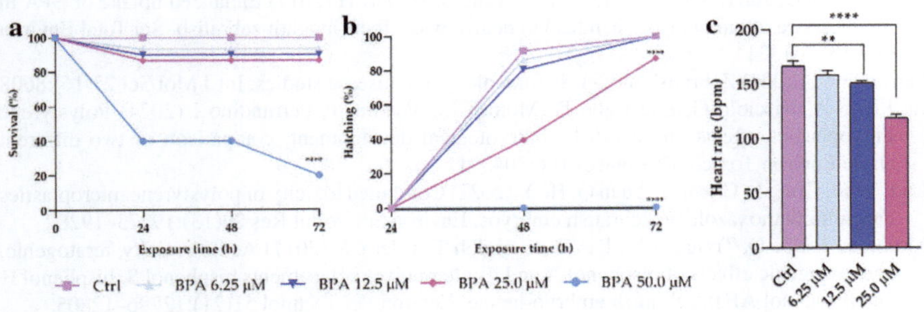

Fig. 23.2 Toxicity parameters during treatment with BPA. **a** Survival; **b** Hatching rate; **c** Heart rate at 72 h of treatment. Dunnett's test (**$p < 0.01$; **** $p < 0.0001$)

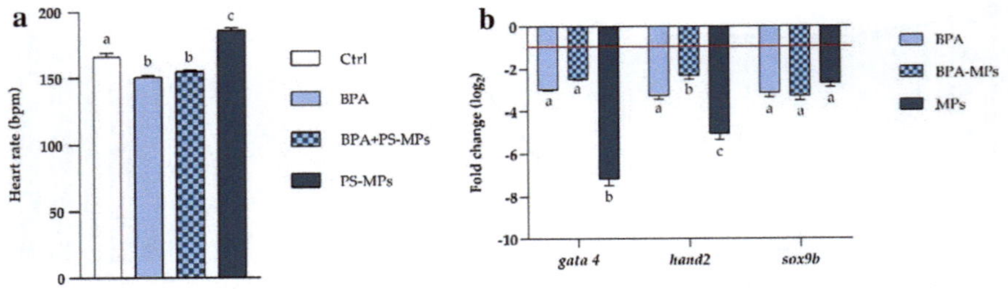

Fig. 23.3 **a** Heart rate at 72 h of co-exposure with BPA 12.5 μM and PS-MPs 1 mgL^{-1}. **b** Fold change of gene expression. Tukey's test: different letters indicate statistically significant differences.

(Fig. 23.3b). Regarding *sox9b*, it was down-regulated in all experimental groups with no significant differences between them (Fig. 23.3b). The heart is the first organ to function and is therefore very sensitive to the effects of environmental pollutants. Furthermore, the action of microplastics in mitigating the effects of BPA has already been demonstrated in other studies in which plastic particles were able to contain the effects of other environmental pollutants by sequestering them (Lu et al. 2022).

Conclusions

In conclusion, both PS-MPs and BPA have been shown to be toxic. During their co-exposure, a slight attenuation of the effects of BPA on the heart was observed, but further studies are needed to better understand their interaction and what happens when their concentrations vary.

References

Chen Q, Yin D, Jia Y, Schiwy S, Legradi J, Yang S, Hollert H (2017) Enhanced uptake of BPA in the presence of nanoplastics can lead to neurotoxic effects in adult zebrafish. Sci Total Environ 609:1312–1321

Ferrandino I (2024) Zebrafish models in toxicology and disease studies. Int J Mol Sci 25(16):8608

La Pietra A, Fasciolo G, Lucariello D, Motta CM, Venditti P, Ferrandino I (2024) Polystyrene microplastics effects on zebrafish embryological development: comparison of two different sizes. Environ Toxicol Pharmacol 106:104371

Lu J, Wu J, Gong L, Cheng Y, Yuan Q, He Y (2022) Combined toxicity of polystyrene microplastics and sulfamethoxazole on zebrafish embryos. Environ Sci Pollut Res 29(13):19273–19282

Moreman J, Lee O, Trznadel M, David A, Kudoh T, Tyler CR (2017) Acute toxicity, teratogenic, and estrogenic effects of bisphenol A and its alternative replacements bisphenol S, bisphenol F, and bisphenol AF in zebrafish embryo-larvae. Environ Sci Technol 51(21):12796–12805

Qin JY, Jia W, Ru S, Xiong JQ, Wang J, Wang W, Hao L, Zhang X (2023) Bisphenols induce cardiotoxicity in zebrafish embryos: role of the thyroid hormone receptor pathway. Aquat Toxicol 254:106354

Rochester JR (2014) Bisphenol A and human health: a review of the literature. Reprod Toxicol 42:132–155

Siddiqui SA, Singh S, Bahmid NA, Shyu DJ, Domínguez R, Lorenzo JM, Pereira JA, Câmara JS (2023) Polystyrene microplastic particles in the food chain: characteristics and toxicity-a review. Sci Total Environ 892:164531

Vancamp P, Houbrechts AM, Darras VM (2019) Insights from zebrafish deficiency models to understand the impact of local thyroid hormone regulator action on early development. Gen Comp Endocrinol 279:45–52

Zhu YY, Liu Y, Xu J, Ni BJ (2022) Three-dimensional excitation-emission matrix (EEM) fluorescence approach to probing the binding interactions of polystyrene microplastics to bisphenol A. J Hazard Mater Adv 5:100046

Chapter 24
Degradation of Biodegradable Polymers in Simulated Marine Environments

Immacolata Liotta, Veronica Ambrogi, Roberto Avolio, Rachele Castaldo, Maria Emanuela Errico, Gennaro Gentile, Federico Olivieri, and Mariacristina Cocca

Introduction

Global plastic production is expected to reach 430 million tonnes in 2024 (Plastic Europe 2025), driven by the material's versatility and low cost (Andrady and Neal 2009). However, the improper end-of-life management has led to significant marine pollution, with millions of tonnes of plastic entering the oceans each year and fragmenting into micro-and nanoplastics (Prata et al. 2019; Santonicola et al. 2025). To mitigate this problem, biodegradable polymers have been proposed (Directive EU 2019). However, some studies show that these materials can persist under natural conditions, releasing microplastics and additives, thus posing similar risks as conventional plastics (Viel et al. 2025; Suaria et al. 2016).

Climate change intensifies the impact of microplastics on the environment, influencing their degradation, distribution, and interaction with ecosystems (Wei et al. 2024). Understanding these impacts is crucial to developing effective strategies to address the microplastic problem in the face of climate change. Rising temperatures can accelerate plastic degradation, resulting in increased fragmentation and the release of smaller microplastic particles (Kamweru et al. 2011).

This study therefore focuses on assessing their degradation under marine-like conditions, considering environmental stressors such as temperature and pH fluctuations, which are expected to intensify with climate change and ocean acidification.

I. Liotta (✉) · R. Avolio · R. Castaldo · M. E. Errico · G. Gentile · F. Olivieri · M. Cocca
Institute of Polymers, Composites and Biomaterials, National Research Council of Italy, Pozzuoli, NA, Italy
e-mail: immacolataliotta@cnr.it

I. Liotta · V. Ambrogi
Department of Chemical, Material and Industrial Production Engineering, University of Naples Federico II, Naples, Italy

G. Gentile et al. (eds.), *Proceedings of the 4th International Conference on Microplastic Pollution in the Mediterranean Sea*, Springer Water,
https://doi.org/10.1007/978-3-032-23847-4_24

Blends of poly(butylene adipate-co-terephthalate) (PBAT), poly(polybutylene succinate) (PBS) and cellulose fibers were aged under abiotic conditions simulating the marine environment by varying temperature and pH to evaluate how these factors influence degradation rates and long-term environmental behavior.

Materials and Methods

Blends containing different compositions of PBAT, PBS and cellulose were prepared: 100 wt% PBAT, 100 wt% PBS, a blend composed by 20 wt% PBAT and 80 wt% PBS (PBAT20-PBS80) without and with the addition of 2.5 phr of cellulose (B2080-CF2.5).

The samples were prepared using a Thermo HAAKE MiniLab twin-screw extruder: about 5 g of material were mixed at 60 rpm for 5 min in the extruder at 160 °C. A degradation test was performed under abiotic conditions reproducing the marine environment for 100 days. Three different experiments were performed, changing temperature and pH of the simulated solutions.

Experiment A was performed at room temperature and pH 8.3, using a 300-watt Osram Ultra Vitalux lamp. Experiment B was performed under the same conditions decreasing pH to 7.3. Experiment C was performed in an Angelantoni Challenge 250E climate chamber at 35 °C, under UV lamp, fixing the pH of the solution to 8.3.

Sample were monitored every month, evaluating their weigh. Furthermore, before and at the end of the ageing period, the surfaces of the samples were analyzed by scanning electron microscopy (SEM) by means of a FEI Quanta 200 FEG SEM (Hillsboro, OR, USA),. The pH of the solutions was also monitored at the beginning and at the end. All the abiotic solutions were filtrated on macroporous silicon filters with 5 μm pore size, MakroPor, purchased from Thermo Fisher Scientific, to collect the microplastics released from the solutions. To identify the chemical nature of the isolated microfragments, MakroPro silicon filters were analysed with a Nicolet iMX 10™ 10 Micro Fourier Transform Infrared Spectroscopy (micro-FTIR) (Thermo Fisher Scientific).

Results and Discussion

After 100 days of exposure, the maximum weight loss of the samples was about 2.5%. Temperature was the main factor influencing degradation, with Experiment C producing the highest weight loss. In particular, the B2080-CF2.5 blend showed a weight loss at 100 days of 2.4 ± 0.8%. Instead, different pH (Experiments A and B) had similar effects. Figure 24.1 presents SEM micrographs of the B2080-CF2.5 sample before and after aging in Experiment A. The presence of heterogeneous cracks and grooves suggests that degradation processes had taken place (Liotta et al. 2025).

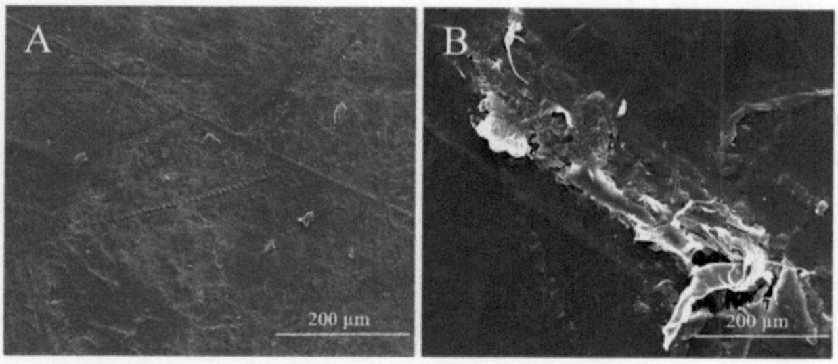

Fig. 24.1 SEM micrographs of the surface of the B2080-CF2.5 sample. A) Unaged sample; B) Sample aged at room temperature and pH 8.3 under a 300-watt Osram Ultra Vitalux lamp (Ageing experiment A)

Table 24.1 pH variation after exposure period in the three experiments

	Exp A	Exp B	Exp C
t0	8.3	7.3	8.3
PBAT	7.2	7.1	7.5
PBAT20-PBS80	7.3	6.9	7.5
B2080-CF2.5	7.2	7.2	7.2
PBS	7.5	7.1	7.6

Polymer degradation also affected the pH of the solution (Table 24.1), causing a notable pH decrease, similar in Experiments A and C.despite the different temperatures. In Experiment B, which started at a lower pH, the recorded pH reduction was less pronounced. These pH changes are consistent with projections for ocean acidification by the end of the century due to anthropogenic emissions (Bindoff et al. 2019). Acidification is likely the result of the release of organic acids and CO_2 during polymer degradation (Romera-Castillo et al. 2023).

FTIR analysis revealed incomplete degradation in several cases, evidenced by the persistence of microplastics (Fig. 24.2) in solutions after aging under lower pH and thermal conditions. These observations raise concerns about secondary pollution and the generation of microplastics, calling into question the assumption that biodegradable plastics provide a sustainable solution and highlighting the complexity of their environmental fate.

Conclusions

This study demonstrates that the environmental performance of biodegradable polyesters is highly variable and strongly influenced by climate stress. Under simulated marine conditions, degradation rates depended on pH, temperature, and material composition, with blends containing PBAT and cellulose showing greater sensitivity

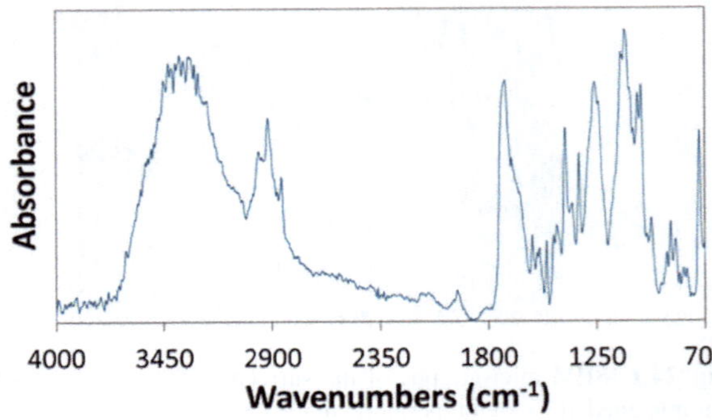

Fig. 24.2 FTIR spectrum of a fragment found in filtrated solution with PBAT20-PBS80 after the exposure period in Exp C

to acidification and thermal stress. Gravimetric and spectroscopic analyses revealed incomplete degradation, including the release of terephthalic acid and microplastics, raising concerns about secondary pollution and microplastic formation. These findings challenge assumptions about the intrinsic sustainability of biodegradable polymers and highlight the need for testing under variable conditions that also reflect climate realities to guide strategies to mitigate plastic pollution.

Acknowledgements This work was performed in the framework of the Project—Biomonitoraggio di micro e nanoplastiche biodegradabili: dall'ambiente all'uomo in una prospettiva one health (BioPlast4Safe)—with the technical and economic support of the Italian Ministry of Health—PNC.

References

Andrady AL, Neal MA (2009) Applications and societal benefits of plastics. Philos Trans R Soc B: Biol Sci 364(1526):1977–1984

Bindoff NL, Cheung WW, Kairo JG, Arístegui J et al (2019) Changing ocean, marine ecosystems, and dependent communities. In: Pörtner HO, Roberts DC et al (eds) IPCC special report on the ocean and cryosphere in a changing climate

Directive (EU) (2019) 2019 of the European Parliament and of the Council of 5 June 2019 on the reduction of the impact of certain plastic products on the environment

Kamweru PK, Ndiritu FG, Kinyanjui TK, Muthui ZW, Ngumbu RG, Odhiambo PM (2011) Study of temperature and UV wavelength range effects on degradation of photo-irradiated polyethylene films using DMA. J Macromol Sci, Part B 50(7):1338–1349

Liotta I, Avolio R, Castaldo R et al (2025) Evaluation of biodegradable plastics as a sustainable solution to marine pollution by microplastics and nanoplastics, proceeding of the 2nd conference on green engineering, sustainable materials, and technologies for circular economy (GREEN CIRC '25)

Plastics Europe (2025) Plastics—the fast facts

Prata JC, Silva ALP, Da Costa JP et al (2019) Solutions and integrated strategies for the control and mitigation of plastic and microplastic pollution. Int J Environ Res Public Health 16(13):2411

Romera-Castillo C, Lucas A, Mallenco-Fornies R, Briones-Rizo M, Calvo E, Pelejero C (2023) Abiotic plastic leaching contributes to ocean acidification. Sci Total Environ 854:158683

Santonicola S, Volgare M, Olivieri F et al (2025) Natural and regenerated cellulosic microfibers dominate anthropogenic particles ingested by commercial fish species from the Adriatic Sea. Foods 14(7):1237
Suaria G, Avio CG, Mineo A et al (2016) The Mediterranean plastic soup: synthetic polymers in Mediterranean surface waters. Sci Rep 6(1):37551
Viel T, Liotta I, Avolio et al (2025) The fate of biodegradable polyesters in the marine environment. Polym Degrad Stab:111539
Wei XF, Yang W, Hedenqvist MS (2024) Plastic pollution amplified by a warming climate. Nat Commun 15(1):2052

Chapter 25
Characterization of the Composition and Types of Marine Microplastics in the Canary Islands

Ana Molina Rodríguez, Miriam Noemí Déniz Martín, Cristina León Santiago, Álvaro Nicolás Cubas Viera, Elsa María Rodríguez-Pérez, Javier Hernández Borges, Eugenio Fraile Nuez, Francisco Machín, and Daura Vega Moreno

Introduction

The Canary Islands, due to their geographical and strategic location in the Atlantic Ocean, as well as the influence of the North Atlantic gyre, have become a hotspots for marine microplastics (MPs). In this area, several studies on microplastics in beaches and in surface waters have estimated that polyethylene (PE) and polypropylene (PP) are the dominant polymers in the coastal areas of the Canary Islands (García-Regalado et al. 2024), being the most abundant in the ocean surface due to their low density (Rodrigues dos Santos et al. 2023). However, although there are studies focused on deep waters (Vega-Moreno et al. 2024), due to the difficulty of sample collection and the complexity of microparticle determination, the information available on their presence, composition and transport is limited. In addition,

A. M. Rodríguez · M. N. D. Martín · C. León Santiago · E. M. Rodríguez-Pérez · D. Vega Moreno
OpenPLAS Group, Chemistry Department, University of Las Palmas de Gran Canaria (ULPGC), Las Palmas de Gran Canaria, Spain

J. Hernández Borges
Department of Chemistry, Departmental Unit of Analytical Chemistry, Faculty of Sciences, University of La Laguna (ULL), San Cristóbal de La Laguna, Spain

E. Fraile Nuez
Oceanographic Center of the Canary Islands, Spanish Institute of Oceanography (IEO), Superior Council of Scientific Research (CSIC) (IEO-CSIC), Santa Cruz de Tenerife, Spain

Á. N. Cubas Viera · F. Machín (✉)
Physical Oceanography and Applied Geophysics (OFYGA), IU-ECOAQUA, University of Las Palmas de Gran Canaria (ULPGC), Las Palmas de Gran Canaria, Spain
e-mail: francisco.machin@ulpgc.es

G. Gentile et al. (eds.), *Proceedings of the 4th International Conference on Microplastic Pollution in the Mediterranean Sea*, Springer Water,
https://doi.org/10.1007/978-3-032-23847-4_25

microplastics are exposed to various processes in the ocean such as weathering processes, photodegradation, chemical oxidation, and physical abrasion, together with the high resistance and permanence of these microplastics, so they undergo fragmentation and degradation, being transported long distances in the ocean (Van Sebille et al. 2020; Veerasingam et al. 2021). Consequently, the high adsorption capacity of microplastics promotes the release of plastic additives into the environment and causes them to accumulate and transport persistent chemical pollutants (POPs) (Abaroa-Pérez et al. 2022). For this reason, microplastics become a double entry vector of pollutants. Therefore, considering that marine currents bring tons of plastics to the coasts of the Canary Islands, the project: "Islas Canarias", subproject 2: "Promoviendo la salud de oceánica de las Islas Canarias" aims to study the marine litter and microplastics that reach the region by evaluating their abundance and composition, in different areas of the islands. For this purpose, the type of microplastic and the composition of the samples collected will be determined using Fourier-transform infrared spectroscopy (FTIR).

Materials and Methods

The sampling campaign was carried out from June 12 to 20, 2025, at 13 selected beaches on the islands of Gran Canaria, Fuerteventura, Lanzarote, and La Graciosa. Table 25.1 presents the names of the beaches, the sampling dates, and the coordinates established. Sampling followed the protocol defined by the Interreg-MAC IMPLAMAC project (https://implamac.com/), with five points sampled at equidistant intervals of 20 m from the high tide line on each beach. At each point, a sampling area was established using 2500 cm^2 (50 × 50 cm) wooden quadrat, and the top 5 cm of sand were sieved and washed with seawater. The sand was then sieved in situ using two mesh sizes (5 and 1 mm), for macroplastics and mesoplastics the first one and microplastic the second one (only MPs larger than 1 mm were analyzed). The collected material was subsequently placed in containers.

In the laboratory, macroplastics, mesoplastics and MPs were subjected to a flotation process using sodium chloride (NaCl). Once decanted, the solution was filtered, and the samples were rinsed with Milli-Q water. Subsequently, the samples were classified according to their size, shape, and colour, with distinctions made between different shades. The polymeric composition was determined using Fourier Transform Infrared Spectroscopy (FTIR), with 10% of the sample analyzed.

Results and Discussion

In total, 2227 plastics particles (macro-, meso-, and microplastics) were quantified: 392 particles in Gran Canaria, 630 particles in Fuerteventura, 496 particles in Lanzarote, and 709 particles in La Graciosa. When classified by size, 1807 were

Table 25.1 Islands, names of the beaches, the sampling dates, and the coordinates established

Island	Beach	Date	Latitude (°N)	Longitude (°W)
Gran Canaria	*Playa de Puerto Rico*	12/06/2025	27.783	15.713
	Playa del Inglés	12/06/2025	27.750	15.568
	Playa de Melenara	13/06/2025	27.987	15.374
	Playa del Hombre	13/06/2025	27.996	15.374
Fuerteventura	*Playa del Matorral*	16/06/2025	28.044	14.333
	Playa Esmeralda	16/06/2025	28.143	14.237
	Grandes Playas	17/06/2025	28.708	13.838
	Playa de las Caletillas	17/06/2025	28.425	13.858
Lanzarote	*Playa de Papagayo*	18/06/2025	28.842	13.788
	Playa de los Pocillos	18/06/2025	28.926	13.635
	Playa de Famara	20/06/2025	29.115	13.556
La Graciosa	*Playa de la Francesa*	19/06/2025	29.220	13.527
	Playa de la Lambra	19/06/2025	29.279	13.495

determined as MPs, 324 as mesoplastics and 96 as macroplastics. These results suggest a predominance of microplastic pollution across all islands, likely due to prolonged degradation of larger plastic items. The notably high count in La Graciosa may be influenced by oceanic currents transporting marine debris from surrounding areas, despite its lower population density. Conversely, the relatively low count in Gran Canaria could reflect localized beach cleaning and the lower influence of ocean currents on the island.

Macro-, Meso- and Microplastic Abundance

Figure 25.1 illustrates the distribution of the total abundance of particles (a) and microplastics (MPs) (b) across the sampled beaches. *Playa de la Lambra* accounted for the highest abundance (709 particles), followed by *Playa de las Caletillas* (585), *Playa de Famara* (492), *Playa de Puerto Rico* (226), and *Playa del Hombre* (121). Lower abundances were recorded at *Playa del Inglés* (44), *Grandes Playas* (26), and *Playa del Matorral* (12). In contrast, only 1 particle was detected at *Playa de Melenara*, 4 particles at *Playa de Papagayo*, and 7 particles at *Playa Esmeralda*. No particles were detected in the sampled areas of *Playa de la Francesa* and *Playa de los Pocillos*. Microplastics followed the same distribution pattern on all beaches. The stark contrast in particle abundance between beaches suggests that local environmental conditions and oceanic current play a significant role in plastic accumulation. Beaches like *Playa de la Lambra, Playa de las Caletillas* and *Playa de Famara*, which are more exposed to prevailing oceanic currents, showed the highest concentrations. In contrast, the absence of particles at *Playa de la Francesa* and *Playa de*

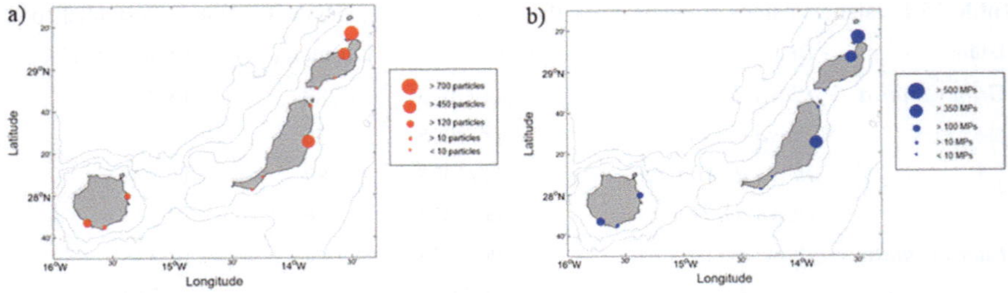

Fig. 25.1 Map of abundance based on the distribution of total marine litter (**a**) and microplastics (MPs) (**b**) sampled in Gran Canaria, Fuerteventura, Lanzarote, and La Graciosa

los Pocillos may reflect effective natural flushing mechanisms or that these are not good accumulation points for microplastic. These findings highlight the need for site-specific management strategies adapted to the exposure and usage patterns of each beach.

Based on plastic type, microplastics were dominated by fragments (91%), followed by fibers (3%), pellets (2%), films (2%), and foams (2%). Among mesoplastics, fragments accounted for 77%, ropes 17%, films 4%, and foams 2%. For macroplastics, ropes were the most abundant (60%), followed by fragments (19%), films (15%), and other forms (6%). Overall, fragments predominated among both microplastics and mesoplastics across all islands, whereas ropes were the dominant category within macroplastics. The predominance of fragments among micro- and mesoplastics suggests ongoing degradation of larger plastic items. The high proportion of ropes in macroplastics may indicate contamination from textiles or marine ropes.

Plastic Composition

A total of 192 samples were analyzed, of which 39 were from Gran Canaria, 65 from Fuerteventura, 18 from Lanzarote, and 70 from La Graciosa. Of these, 91% of were identified as polyethylene (PE) and 9% as polypropylene (PP). In Gran Canaria, PE accounted for 95% and PP for 5%; in Fuerteventura, PE was 92% and PP 8%; in Lanzarote, PE was 89% and PP 11%; and in La Graciosa, PE was also 89% and PP 11%. The dominance of polyethylene (PE) on all islands is consistent with previous studies such as Edo et al. (2019). Both polymers are known for their persistence in marine environments, reinforcing the need for upstream interventions to reduce plastic input into coastal systems.

Conclusions

The study revealed widespread plastic contamination across 13 beaches in the Canary Islands, with microplastics being the most abundant (81%). *Playa de la Lambra* showed the highest particle count, while several beaches had minimal or no detected plastics, such as *Playa de la Francesa* and *Playa de los Pocillos.* Fragments dominated among micro- and mesoplastics, whereas ropes were most common in macroplastics. Polyethylene (PE) was the predominant polymer (91%), followed by polypropylene (PP) at 9%, with consistent distribution across all islands. These results highlight the need for continuous monitoring and the creation of a specific protocol for oceanic islands.

References

Abaroa-Pérez B, Ortiz-Montosa S, Hernández-Brito JJ, Vega-Moreno D (2022) Yellowing, weathering and degradation of marine pellets and their influence on the adsorption of chemical pollutants. Polymers (Basel) 14. https://doi.org/10.3390/polym14071305

Edo C, Tamayo-Belda M, Martínez-Campos S et al (2019) Occurrence and identification of microplastics along a beach in the Biosphere Reserve of Lanzarote. Mar Pollut Bull 143:220–227. https://doi.org/10.1016/j.marpolbul.2019.04.061

García-Regalado A, Herrera A, Almeda R (2024) Microplastic and mesoplastic pollution in surface waters and beaches of the Canary Islands: A review. Mar Pollut Bull 201

Rodrigues dos Santos C, Pinheiro Drumond G, Rezende Moreira V, et al (2023) Microplastics in surface water: occurrence, ecological implications, quantification methods and remediation technologies. Chem Eng J 474

Van Sebille E, Aliani S, Law KL, et al (2020) The physical oceanography of the transport of floating marine debris. Environ Res Lett 15

Veerasingam S, Ranjani M, Venkatachalapathy R et al (2021) Contributions of Fourier transform infrared spectroscopy in microplastic pollution research: a review. Crit Rev Environ Sci Technol 51:2681–2743. https://doi.org/10.1080/10643389.2020.1807450

Vega-Moreno D, Sicilia-González S, Domínguez-Hernández C, et al (2024) Exploring the origin and fate of surface and sub-surface marine microplastics in the Canary Islands region. Front Mar Sci 11. https://doi.org/10.3389/fmars.2024.1314754

Chapter 26
Abiotic Degradation of Biodegradable Plastics in Water: Chemical Characterization and Ecotoxicological Effects on *Daphnia Magna*

Chiara Magnabosco, Matteo Saffioti, Monica Passananti, Roberta Piscia, Silvia Galafassi, and Serena Zaccara

Introduction

To reduce the ecological impacts of plastic pollution, the use of biodegradable plastics (BDPs) is increasingly spreading as an alternative to conventional plastics (Campanale et al. 2024; Capolupo et al. 2023) with global production expected to grow from 1 million tons in 2023 to 5 million tons by 2028 (Bioplastic market development update, 2023). BDPs are defined as material capable of being decomposed by microorganisms into water, biomass and carbon dioxide (Kim et al. 2023). However, their degradation strongly depends on both material properties and environmental conditions, which are influenced by abiotic and biotic factors (Afshar et al. 2024; Ribba et al. 2022). In natural environments, degradation is often incomplete,

C. Magnabosco (✉) · R. Piscia · S. Galafassi
National Research Council, Water Research Institute, Verbania (VB), Italy
e-mail: chiaramagnabosco@cnr.it

S. Galafassi
e-mail: silvia.galafassi@cnr.it

C. Magnabosco · S. Zaccara
Department of Theoretical and Applied Sciences, University of Insubria, Varese (VA), Italy

M. Saffioti · M. Passananti
Department of Chemistry, University of Torino, Torino (TO), Italy
e-mail: monica.passananti@unito.it

M. Passananti
Institute for Atmospheric and Earth System Research, University of Helsinki, Helsinki, Finland

S. Galafassi
National Biodiversity Future Center, Piazza Marina, Palermo (PA), Italy

G. Gentile et al. (eds.), *Proceedings of the 4th International Conference on Microplastic Pollution in the Mediterranean Sea*, Springer Water,
https://doi.org/10.1007/978-3-032-23847-4_26

potentially leading to the release of intermediate degradation products (BDP-DPs), including additives, potential toxic compounds, monomers and microplastics (Afshar et al. 2024; Cera et al. 2020). This study investigates the impact of BDPs on freshwater environments: abiotic degradation tests were conducted under controlled conditions, and the resulting leachates were analyzed to identify chemical compounds. Toxicity assays were then performed to assess potential impacts on aquatic organisms.

Materials and Methods

Abiotic degradation experiments consisted of exposing 3 mg/L of BDP pellets to UV radiation for two weeks. Three pure polymers were selected—polycaprolactone (PCL), polybutylene adipate terephthalate (PBAT) and polylactic acid (PLA)—and immersed in pure water, with dark-incubated samples serving as controls. The same experimental setup was also applied to pellets immersed in water collected from Lake Maggiore and Lake Varese. The resulting BDP-DPs were analysed by Anion Exchange Chromatography and Ultra High-Performance Liquid Chromatography coupled with an Orbitrap analyser. Total carbon content (TC) was measured with a TOC analyser, and structural changes in the pellets were assessed using Fourier Transform Infrared Spectroscopy (FTIR). Pellet weight loss was also recorded. Finally, acute standard toxicity tests on *Daphnia magna* were performed with leachates obtained from lake water, using the commercial Daphtoxkit F kit (Ecotox).

Results and Discussion

Physicochemical Analysis

Physicochemical analysis of BDP-DPs included total carbon release (TC), ionic content and mass spectrometry. FTIR analysis and weight measurements were performed on the incubated pellets. Preliminary results (Fig. 26.1) showed that PCL released the highest amount of TC, followed by PBAT, across all leachates, while PLA didn't release significant amount of TC. Leachates from pure water showed the highest TC values, followed by Lake Varese and then Lake Maggiore. Except for PCL in pure water, UV-exposed samples released more carbon than their respective dark-incubated controls. Ionic analysis revealed presence of different ions, including fluoride, lactate, glycolate, acetate, 6-hydroxyhexanoate, propionate, formate, chloride, nitrite, hexanoate, sorbate, nitrate, adipate, malate, sulphate, oxalate, and terephthalate. No consistent trend was observed, indicating possible dynamic degradation processes and compound transformation over time. Untargeted analysis performed by UHPLC-MS/MS in pure water confirmed the results obtained through TC and IC analysis: PLA showed very high stability, releasing in water almost exclusively

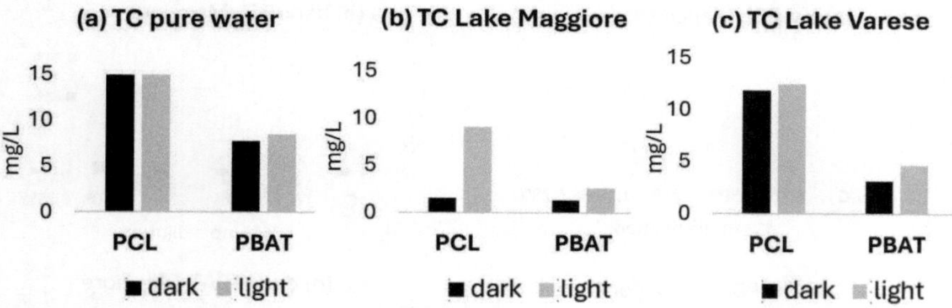

Fig. 26.1 Concentration of total carbon release in PCL and PBAT leachates obtained with pure water (**a**), Lake Maggiore water (**b**) and Lake Varese water (**c**) after 14 days of UV-exposition with corresponding control samples kept in the dark

surfactants and traces of lactate. PCL and PBAT, on the other hand, were less stable, releasing additives and various DPs. All the monomers and other molecules, such as bis(4-hydroxybutyl) adipate, that derived directly from the degradation of the polymer chain were observed. FTIR spectra showed structural changes only in PBAT pellets, with a new peak at 1724 cm^{-1}, associated with $C = O$ stretching in ester bonds, likely reflecting hydration effects. No significant weight loss was observed in any incubated pellets. However, FTIR observations of oxygenated functions indicate that polymer degradation induces chemical changes on the pellets' surface. The latter could influence the interaction between BDPs and microorganisms or other chemical species.

Toxicity Effects on Daphnia Magna

As summarized in Fig. 26.2, no significant acute toxic effects were observed. For Lake Maggiore leachates, PCL exhibited a maximum effect of 15% in UV-exposed samples and 10% in dark-incubated samples. PBAT showed no notable difference from the control, with both control and treated samples showing a 10% effect—within the threshold considered acceptable for test validity. PLA leachates showed no observable effects (0%). Similarly, leachates from Lake Varese revealed no significant toxicity. PCL reached a maximum of 10% effect in dark condition, while the UV-exposed sample induced 5% effect. PBAT leachates reached up to 10% effect in several dilutions under UV exposure. Notably, in Lake Varese samples, some *Daphnia* neonates exhibited apparent moulting inhibition, particularly at the 6.25% dilution of dark-incubated PCL leachate and across several UV-exposed PBAT leachate dilutions. However, no concentration–response relationship was observed in any test, suggesting an overall absence of acute toxicity in *Daphnia magna*.

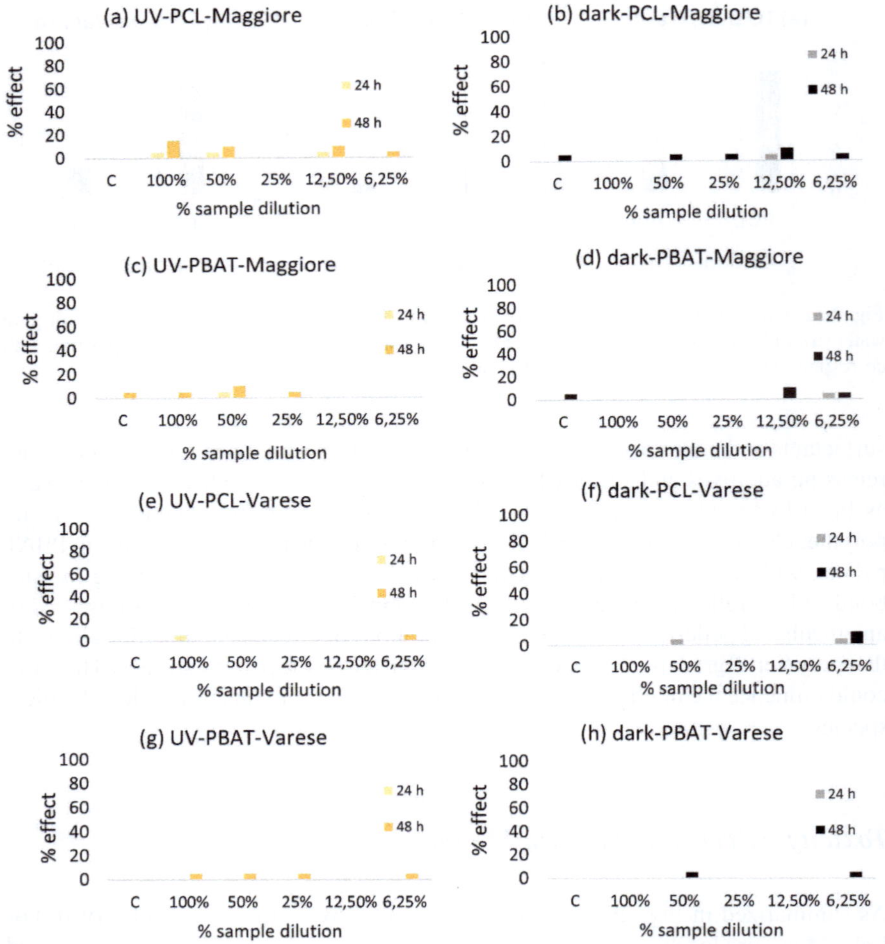

Fig. 26.2 Percentage of effect at each dilution after 24 and 48 h of exposure for UV-exposed and dark control samples: PCL leachates prepared with Lake Maggiore (**a**, **b**) and Lake Varese (**e**, **f**) water; PBAT leachates with Lake Maggiore (**c**, **d**) and Lake Varese (**g**, **h**) water. UV-exposed samples and corresponding dark controls are shown for each material

Conclusions

Preliminary results about abiotic degradation experiments on selected BDPs indicate only minor physicochemical alterations in the incubated pellets. The chemical analysis of leachates revealed that TC release was influenced by both the polymer type and the incubation medium (pure water vs. lake water), suggesting potential interactions between BDPs and the chemical composition of natural waters. Mass spectrometry analysis and variations in ionic concentrations may reflect the formation of new chemical compounds and ongoing degradation processes. No acute toxicity effects of obtained leachates were recorded on *Daphnia magna*. However, moulting effects observed in leachates obtained with Lake Varese water suggest potential sublethal effects.

Acknowledgement This study was funded by the European Union - NextGeneration EU, Mission 4 Component 2 Investment 1.1 CUP 2022EW9CZN project PRIN BIOPLACE. The views and opinions expressed are solely those of the authors and do not necessarily reflect those of the European Union, nor can the European Union be held responsible for them.

References

Afshar SV, Boldrin A, Astrup TF, Daugaard AE, Hartmann NB (2024) Degradation of biodegradable plastics in waste management systems and the open environment: a critical review. J Clean Prod 140000

Bioplastics market development update 2023. https://www.european-bioplastics.org/bioplastics-market-development-update-2023-2/

Campanale C, Galafassi S, Di Pippo F, Pojar I, Massarelli C, Uricchio VF (2024) A critical review of biodegradable plastic mulch films in agriculture: definitions, scientific background and potential impacts. Trends Anal Chem 170:117391

Capolupo M, Rafiq A, Coralli I, Alessandro T, Valbonesi P, Fabbri D, Fabbri D (2023) Bioplastic leachates characterization and impacts on early larval stages and adult mussel cellular, biochemical and physiological responses. Environ Pollut 319:120951

Cera A, Cesarini G, Spani F, Scalici M (2020) Hydra vulgaris assay as environmental assessment tool for ecotoxicology in freshwaters: a review. Mar Freshw Res 72(6):745–753

Kim MS, Chang H, Zheng L, Yan Q, Pfleger BF, Klier J, Nelson K, Majumder ELW, Huber GW (2023) A review of biodegradable plastics: chemistry, applications, properties, and future research needs. Chem Rev 123(16):9915–9939

Ribba L, Lopretti M, de Oca-Vásquez GM, Batista D, Goyanes S, Vega-Baudrit JR (2022) Biodegradable plastics in aquatic ecosystems: latest findings, research gaps, and recommendations. Environ Res Lett 17(3):033003

Chapter 27
Impact of PET Nanoplastics on Immune Homeostasis and Cellular Function in *Mytilus galloprovincialis*

Lucia Therese Marcianò, Claudia La Corte, Mariele Staropoli, Federica Bertini, Luca Bisanti, Mariano Dara, Matteo Cammarata, and Maria Giovanna Parisi

Introduction

The innate immune system of marine organisms, particularly mussels such as *Mytilus galloprovincialis*, encompasses a sophisticated system of responses and defence mechanisms against various stressors. Owing to their sessile and filter-feeding lifestyle, mussels are recognized as sentinel species for monitoring environmental contamination, with their immune responses acting as early warning biomarkers of ecosystem health (Parisi et al. 2021). Environmental contaminants can significantly impair the mussel immune system, leading to increased susceptibility to disease and widespread physiological dysregulation (Bertini et al. 2025). The small size (<100 nm) and high bioavailability of nanoplastics allow direct interactions with hemocytes and the activation of immune responses. In this context, the present study applied a multi-level assessment combining in vivo and in vitro approaches to investigate the effects of PET nanoparticles (PET NPs), as polyethylene terephthalate (PET) is a plastic polymer commonly detected in coastal environments and mussels' tissues (Fraissinet et al. 2024).

L. T. Marcianò
Department of Earth and Marine Sciences, Palermo, Italy

L. T. Marcianò · C. La Corte · M. Staropoli · F. Bertini · L. Bisanti · M. Dara · M. Cammarata · M. G. Parisi (✉)
Marine Immunobiology Laboratory, Department of Earth and Marine Sciences (DiSTeM), University of Palermo, Palermo, Italy
e-mail: mariagiovanna.parisi@unipa.it

G. Gentile et al. (eds.), *Proceedings of the 4th International Conference on Microplastic Pollution in the Mediterranean Sea*, Springer Water,
https://doi.org/10.1007/978-3-032-23847-4_27

Materials and Methods

Mytilus galloprovincialis specimens for both *in vivo* and *in vitro* experiments were purchased from a local farmer and acclimated for seven days in artificial seawater prepared with Haquoss Hi-Tech Reef salt dissolved in freshwater, salinity 37–38 ± 1‰, 15 ± 1 °C, 12:12 h light/dark cycle at the Marine Immunobiology Laboratory, University of Palermo, Italy. For *in vivo* exposure, after acclimation, mussels were randomly allocated to four 20 L aquaria per treatment, with continuous water mixing, and exposed for seven days to PET NPs (<100 nm; stock solution 1 mg/mL in distilled water) at concentrations of 0 (control), 0.05, 0.5, and 1 mg/L. Gill and digestive gland samples were collected from three random individuals after 24 h, 48 h, 72 h, and 7 days from the beginning of exposure, with each treatment performed in triplicate. Tissues were homogenized in TBS, centrifuged (15.000 rpm, 20 min, 4 °C), and the supernatants were analysed for protein concentration (Bradford 1976), and extracts were standardized to 0.5 mg/mL. For *in vitro* exposure, after the acclimatation period, hemolymph (~1 mL per mussel) was withdrawn from the posterior adductor muscle of five randomly selected individuals using sterile 1 mL syringes. Samples from five individuals were pooled to obtain one replicate, and four independent pools were generated for each treatment group. Each pool was then exposed to PET NPs at concentrations comparable to those used in vivo studies. All treatments were performed in triplicate.

Biochemical, Protein and Cellular Assays in Mussels' Tissue and Hemocytes After PET NPs Exposure

Following *in vivo* exposure, biochemical and protein analyses were performed on mussel gill and digestive gland extracts. To measure ***Glutathione Peroxidase (GPx) activity***, digestive gland and gill extracts (50 μL) were incubated with 100 μL TMB (Sigma-Aldrich) in 96-well plates for 30 min in the dark; the reaction was stopped with 50 μL 2 M H_2SO_4 and absorbance was measured at 450 nm (RAYTO RT-2100C). Enzyme activity was expressed as U/mg protein (Parisi et al. 2021). For ***Lysozyme (LYS) activity***, 15 μL samples were incubated in triplicate with 250 μL *Micrococcus lysodeikticus* suspension in 96-well plates at 37 °C for 20 min, with TBS as a blank. The absorbance at 450 nm was measured before and after the incubation. One unit was defined as 0.001 absorbance reduction/min (Sutton et al. 2006). ***Phenoloxidase (PO) activity*** was determined spectrophotometrically using L-DOPA (Sigma-Aldrich). Samples (50 μL) were incubated with 50 μL trypsin (1 mg/mL) or water (blank) in a reaction mixture containing 5 mM L-DOPA and 20.7 mM MBTH for 20 min at 20 °C. The absorbance at 490 nm was recorded after 20 and 60 min. Activity was expressed in U/min (1 U = 0.001 ΔA_{490} min^{-1} mg^{-1} protein)

(Winder and Harris 1991). ***Western blotting*** was performed to verify antibody specificity in the digestive gland and gill tissue extracts using a mouse monoclonal anti-HSC70 (SAB3701436; Sigma-Aldrich). Proteins were separated on SDS-PAGE (4% stacking, 10% polyacrylamide resolving gel; 30 min at 180 V, Bio-Rad mini gel kit) and transferred to nitrocellulose membranes with a semi-dry blotting system (Bio-Rad Laboratories; 75 min at 0,8 mA/cm^2 in blotting buffer, 20 mM Tris–HCl, 192 mM glycine, 20% methanol, pH 8.8). After blocking in TBS (3% BSA, 1% Tween-20, 1 h), the membranes were incubated overnight with anti-HSC70 (1:1000) in blocking solution, washed three times with blocking buffer, and incubated for 1 h with anti-mouse IgG-AP (1:5000 in washing buffer, 0.1% BSA; A3562). The signal was developed using BCIP/NBT staining. For HSC70 modulation, ***Dot-blot analysis*** was performed to determine. Tissue extracts (30 μg) were applied to nitrocellulose membranes (Bio-Dot Apparatus, Bio-Rad) with the same antibody dilutions as in the western blotting. Membranes were pre-washed in TBS (pH 7.4; 1.21 g Trizma base, 9 g NaCl), samples filtered by gravity (1 h), blocked in TBS + 1% BSA + 0.05% Tween-20 (1 h), and washed 3 × with 0.1% TBS-T. Incubation was performed with the primary antibody (1 h, TBS-T + 0.1% BSA), followed by three washes and incubation with a secondary antibody conjugated to alkaline phosphatase (1 h). After washing (2 × TBS-T; 1 × TBS), the membranes were developed using BCIP/NBT. Densitometric analysis was performed using Fiji software; Integrated Density Values (IDV) were calibrated and background-subtracted for normalized comparisons.

After *in vitro* exposure, hemocyte responses were evaluated. ***Cell viability*** was assessed using the trypan blue exclusion test (0.01% in phosphate-buffered saline). Hemocyte suspension (90 μL) was mixed with 10 μL Trypan blue and examined under a light microscope (20 ×, Leica DM750) taking photos every 5 min for 30 min to determine their viability. Non-viable cells were stained blue and counted against the total number of cells. ***Motility*** was assessed by imaging hemocytes exposed to different PET NP concentrations every 5 min for 3 h using a light microscope (Leica DM750) with ORMA-Eurotek MDH5 camera (Milan, Italy). Cells were first tracked with the Manual Tracker plugin in ImageJ and subsequently analyzed with the Chemotaxis and Migration Tool software (Ibidi). All experiments were performed in triplicates.

Results and Discussion

Regarding enzymatic responses (Fig. 27.1a–c), the initial increase in glutathione peroxidase (GPx) activity at low concentrations particularly in the gill tissue may likely represents an early antioxidant response against PET NP-induced oxidative stress, as glutathione peroxidase detoxifies hydrogen peroxide and lipid hydroperoxides. However, the subsequent decline at higher concentrations and longer exposure durations suggests the depletion or inhibition of the antioxidant system. The marked reduction in lysozyme (LYS) activity in the digestive gland indicates suppression of antimicrobial defence, as lysozyme contributes to bacterial clearance by degrading

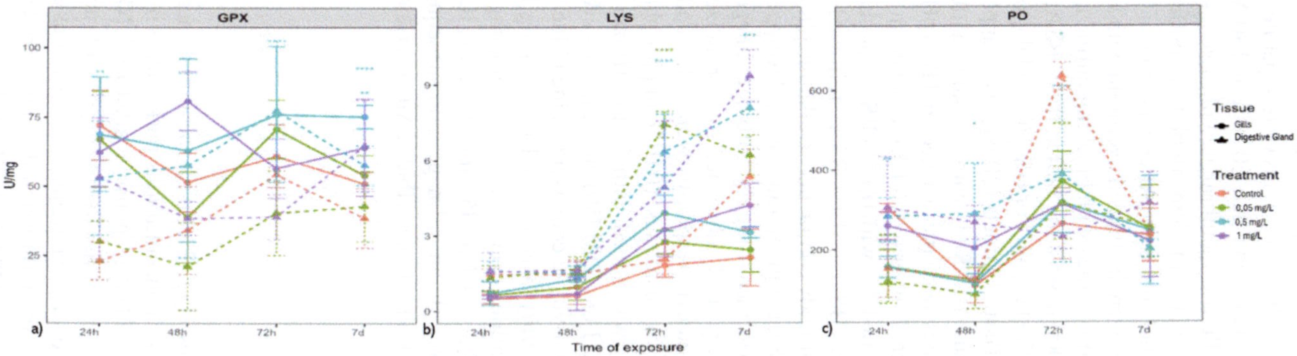

Fig. 27.1 Temporal trends of enzymatic activity in *Mytilus galloprovincialis*. Statistical significance compared to control (*$p < 0.05$; ** $p < 0.01$; *** $p < 0.001$; **** $p < 0.0001$) (**a**) Glutathione Peroxidase (GPx) (**b**) Lisozyme (LYS), and (**c**) Phenoloxidase (PO) activity in gills and digestive glands

peptidoglycan. Similarly, the transient peaks of phenoloxidase (PO) in the gills, followed by its inhibition, point to an early but not sustained activation of the proPO system, central to melanization and encapsulation responses. These trends suggest an initial mobilization of defence pathways, followed by functional impairment due to chronic stress.

This dual pattern is consistent with the biphasic modulation of HSC70, molecular chaperone involved in protein folding and stress recovery and highly conserved across invertebrates, highlighting its role in both acute and long-term adaptive response contributing to resilience against environmental challenges (La Corte et al., 2025) (Fig. 27.2).

The early downregulation after acute exposure may reflect stress-induced destabilization of protein homeostasis, while later upregulation reflects a compensatory mechanism aimed at restoring normal cellular function during prolonged exposure to stress.

The significant reduction in cell viability and motility at higher concentrations suggests that NPs compromise hemocyte functionality, potentially impairing immune competence (Fig. 27.3a). The stronger inhibition of motility at 0.5–1 mg/L PET NPs supports the hypothesis that NPs interfere with migratory processes, essential for immune surveillance and wound healing in bivalves (Fig. 27.3b, c). These impairments suggest that PET NPs compromise immediate cellular defences, ultimately reducing host fitness and increasing susceptibility to pathogens.

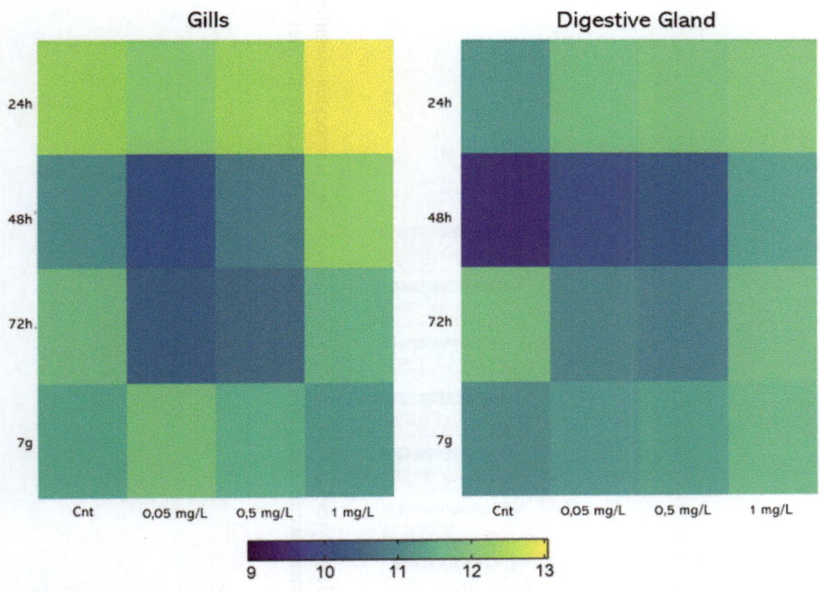

Fig. 27.2 Heat map of HSC70 protein modulation highlighting variations across samples

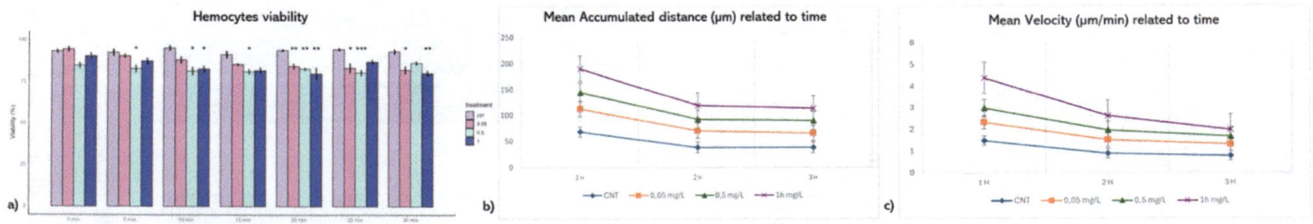

Fig. 27.3 Hemocytes viability and motility analysis. (**a**) Hemocytes viability (%), asterisks indicate statistical significance (*$p < 0.05$, **$p < 0.01$, ***$p < 0.001$); (**b**) mean accumulated distance (μm) over time; (**c**) mean hemocytes velocity (μm/min) over time

Conclusions

The alterations in immune responses induced by PET NPs in *Mytilus galloprovincialis* as initial activation of protective pathways followed by the suppression of immune and antioxidant functions, may undermine mussel health and resilience, ultimately affecting their ecological role as sentinel species and posing risks for ecosystem stability.

References

Bertini F, Catania V, Calabrisotto LS, Dara M, Bisanti L, La Corte C, Cammarata M (2025) A multi-comprehensive approach to assess the responses of the Mediterranean mussel *Mytilus galloprovincialis* (Lamarck, 1919) to a simulation of a diesel-oil mixture spill. Aquat Toxicol 279:107188

Bradford MM (1976) A rapid and sensitive method for the quantitation of microgram quantities of protein utilizing the principle of protein-dye binding. Anal Biochem 72(1–2):248–254

Fraissinet S, De Benedetto GE, Malitesta C, Holzinger R, Materić D (2024) Microplastics and nanoplastics size distribution in farmed mussel tissues. Commun Earth Environ 5(1):128

La Corte C, Barnay-Verdier S, Furla P, Bisanti L, Dara M, Rizzuto G, Cammarata M (2025) Integrated approach to explore *Anemonia viridis* regeneration under a climate change scenario. Sci Rep 15(1):25298

Parisi MG, Pirrera J, La Corte C et al (2021) Effects of organic mercury on *Mytilus galloprovincialis* hemocyte function and morphology. J Comp Physiol B 191:143–158

Sutton J, Balfry S, Higgs D, Huang CH, Skura B (2006) Impact of iron-catalyzed dietary lipid peroxidation on growth performance, general health and flesh proximate and fatty acid composition of Atlantic salmon (*Salmo salar L.*) reared in seawater. Aquaculture 257(1–4): 534–557

Winder AJ, Harris H (1991) New assays for the tyrosine hydroxylase and dopa oxidase activities of tyrosinase. Eur J Biochem 198(2):317–326

Chapter 28
Environmental Impact of Fibrous Microplastics Emitted from Textiles During Washing Under the Life Cycle Assessment Framework

Massimiliano Mariani, Raffaella Mossotti, Elisabetta Abbate, Tiziano Battistini, and Carlo Brondi

Abstract The massive employment of plastics in various industrial sectors has led to an uncontrolled increase and mismanagement of plastic waste, contributing to the rampant presence of plastics in the environment. Primary and secondary microplastics are a significant environmental and health concern due to persistence and ability to cross biological barriers. Additionally, microplastics are an excellent vehicle for chemical additives and pollutants. Microplastics represent a concerning environmental problem currently not addressed by the Life Cycle Assessment (LCA) methodology, the most common tool in evaluating the environmental impact of products, services, and activities. In recent years, several works have emerged to present a scientific approach to the quantification of the unitary environmental effects of microplastics through existing and novel impact categories. In this project, a cohesive procedure was established, based on ISO 4484 series, for the simulation of a washing

M. Mariani (✉) · C. Brondi
Institute of Intelligent Industrial Technologies and Systems for Advanced Manufacturing (STIIMA), National Research Council of Italy (CNR), Milano, Italy
e-mail: massimilianomariani1@cnr.it

C. Brondi
e-mail: carlo.brondi@stiima.cnr.it

R. Mossotti
Institute of Intelligent Industrial Technologies and Systems for Advanced Manufacturing (STIIMA), National Research Council of Italy (CNR), Biella, Italy
e-mail: raffaella.mossotti@stiima.cnr.it

E. Abbate
Department of Environmental Science, Radboud Institute for Biological and Environmental Sciences (RIBES), Radboud University, Nijmegen, The Netherlands
e-mail: elisabetta.abbate@ru.nl

T. Battistini
Aquafil S.p.A., Trento, Italy
e-mail: tiziano.battistini@aquafil.com

G. Gentile et al. (eds.), *Proceedings of the 4th International Conference on Microplastic Pollution in the Mediterranean Sea*, Springer Water,
https://doi.org/10.1007/978-3-032-23847-4_28

process, collection and filtration of effluents on silicon filters, and optical and chemical classification of emitted fibrous microplastics through μ-FT-IR spectroscopy. Spectroscopy results were compared to a gravimetric analysis. The spectroscopy analysis of PA6, PA6.6, PET, and PP samples revealed that fibrous microplastics of the same chemical species as the sample are equal to or less than 0.3% of the total emitted matter, confirming an extensive organic and synthetic material contamination. The quantity of emitted microfilaments doesn't significantly change between the first and third wash. An environmental assessment through the impact category "Physical effects on biota", revealed that the direct emissions to the marine compartment of PET microfilaments from the polyethylene terephthalate sample have the largest impact at the midpoint and endpoint level, equal to 9.72 E-01 PAF m3 day/kg_{sample} and 2.65 E-05 PDF m2 year/kg_{sample}.

Introduction

Plastics have become an indispensable resource in the global manufacturing industry due to their valuable qualities. However, their widespread use raises critical environmental concerns, as the persistence and fragmentation of plastic products, including synthetic textiles, contribute to the generation of microplastic pollutants that enter wastewater streams during washing, accumulate in ecosystems throughout the product life cycle (Browne et al. 2011), and, once discarded, may continue to represent a source of microplastics. Life Cycle Assessment (LCA) is one of the most adopted scientific methods to quantify the environmental impact of a product along its life cycle, defined in ISO 14040 and 14044 standards. In LCA, impacts are calculated by mapping the emissions into the environment and subsequently by associating each emission with a set of factors, converting them into multiple impact categories. Life cycle inventories and characterization factors are fundamental to thoroughly account for a product's environmental burden. To date, few methodologies support a science-based approach to include microplastics in LCA. (Quantis 2020; Corella-Puertas et al. 2023; Schwarz et al. 2023). The international working group MariLCA addressed the concern of microplastics in the aquatic environment with the introduction of a novel impact category called "*Physical effects on the biota*", which aims to describe the potential effect on marine species due to internal and external pathways (Saadi et al. 2025). While this approach doesn't cover the chemical impacts of plastic and plastic additives, it can provide valuable insights into the effects of ingestion, entanglement, and smothering of plastics. In this work, a science-based method was developed to experimentally evaluate the fibrous microplastic emissions during washing under controlled standard conditions. A rigorous and consensus-based approach was followed for the simulation of washing conditions. Adaptations were applied to harmonize the washing and characterization procedure. The optical analysis and FTIR spectroscopy were carried out per UNI ISO 4484-2 regulation, allowing for a physico-chemical description of the released microplastic particles. The environmental impact of fibrous microplastics on marine water and sediments was assessed thanks to the MarILCA method, providing a detailed description of the physical effects of each microplastic type on the aquatic biota populating the receiving compartment.

Materials and Methods

Four knitted, rectangular, tubular 24 cm × 10 cm specimens, without seams (Dalla Fontana et al. 2021), of different synthetic fabrics were tested: polyamide 6.6 (PA6.6), polyamide 6 (PA6), polyethylene terephthalate (PET), and polypropylene (PP). Fabrics were supplied by our partner Aquafil S.p.A. The characteristics of the yarn fibers are reported in Table 28.1. All fabrics are produced from continuous filaments opacified with TiO_2. The identity of each fabric type was previously confirmed by FTIR spectroscopy. The fabrics were washed following the conditions described in EN ISO 4484-1, and the Gyrowash James Heal model 16158 device was used. None of the plain samples were subjected to pre-washing or scouring treatments. For each sample, three repetitions have been performed. The wastewater from the 1st wash and the 3rd wash was collected. The obtained suspension was filtered through a silicon filter mounted on a glass filtration apparatus connected to a vacuum pump. A micro-porous silicon membrane with the following characteristics: size 13 mm × 13 mm, pore size 5 μm, square pore geometry, internal distance 12 μm, and thickness 500 μm was used. Filters were subjected to a cleaning sonication procedure to remove any particles derived from packaging. Silicon filters were chosen due to their transparency to the broad midinfrared region of 4000–600 cm^{-1}, and their compatibility in identifying polyethylene and polypropylene (Käppler et al. 2015). Filters were kept overnight in a vacuum oven at 60 °C to remove moisture. To reduce cross-contamination, canisters and steel balls were cleaned before each test. The material loss was first evaluated through gravimetric analysis and then compared to μ-FT-IR results. The filter surface was analyzed using a liquid nitrogen MCT detector μ-FT-IR Nicolet iN 10 infrared microscope and processed using the proprietary image analysis tool OMNIC PICTA. All the collected particles were selected using the Particle Wizard section with the following operative conditions: 24 co-scans (100 × 100 μm), spectral range between 4000 and 650 cm^{-1}, spectral resolution of 4 cm^{-1}. Plastic particles were distinguished from the non-plastic ones, such as cellulose and protein material, by comparison to standard libraries with a minimum HQI threshold of 0.7. Manual intervention by the operator was still necessary during data processing. The microfilaments' morphology and size (length and width) were measured by optical microscopy, Leica DMLP. The potential environmental impact on the organisms populating marine water and sediments was assessed through the novel impact category "*Physical effects on biota*". The declared unit has been defined as a kilogram of a dry knitted sample 100% synthetic. The system boundaries are restricted to the Use Phase of the products, assuming that the only relevant impact pathway is wastewater produced during washing. No allocation procedures have been applied.

Table 28.1 Characteristics of yarn fibers composing the fabrics

Fiber Characteristics	PA6.6	PA6	PET	PP	Norm
Average yarn count (dtex)	44	44	55	52	ISO 2060, 2062
Average tenacity (cN/dtex)	3.9	4.3	4.1	3.9	ISO 2062
N° filaments	34	44	72	42	–

Results and Discussion

In Fig. 28.1, the material loss is reported as evaluated by gravimetric analysis. Results show a general reduction in material loss after the first wash, except for PA6 samples. The absence of significant differences between samples was confirmed by the Tukey method. The gravimetric result was compared to the material loss identification by FTIR spectroscopy. Table 28.2 reports the average count of microparticle loss and standard deviations after one and three washes.

A precise analysis of the PP fragments revealed that only 2 particles are compatible with a fibrous shape; therefore, the average fibrous microplastic emissions are equal for all samples, independently of wash repetitions and the number of filaments composing the fibers. No significant dimensional difference was highlighted between the first and third washes. Fibers show a diameter between 8 and 15 μm and an average length of around 500 μm. Moreover, the material loss from the PA sample is primarily related to organic contaminants since fibrous microplastics account for less than 0.3% of the total emission, highlighting the unreliability of gravimetric results as a tool to quantify the microplastic discharges. Fibrous and non-fibrous synthetic pollutants, not traceable to the original samples, account for up to 2.8% of the overall emissions. Contaminations from other sources, such as packaging and other products, account for a significant share of the emissions. The presence of

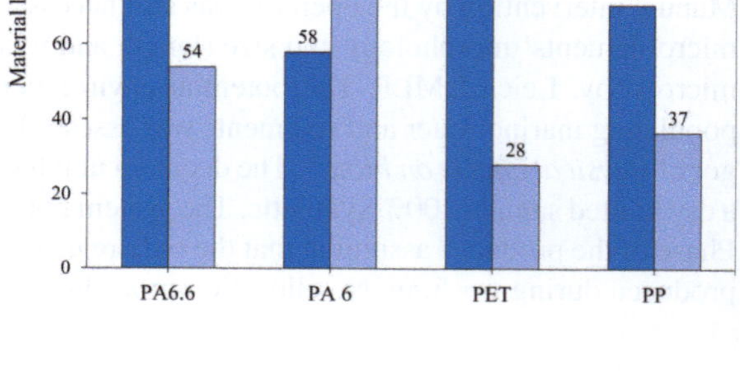

Fig. 28.1 Average material loss after the first and third wash evaluated by gravimetric analysis

Table 28.2 Count of average microparticles loss after the 1st and 3rd wash by FTIR spectroscopy

Material	Wash	Natural		Synthetic	
		Cellulose	Other	Sample	Others
PET	1	223 ± 63	305 ± 329	1.33 ± 0.58	22 ± 14
	3	12 ± 13	115 ± 125	1.5 ± 0.70	1 ± 0.71
PA6	1	245 ± 98	154 ± 42	1.50 ± 0.57	11 ± 5
	3	0	759 ± 64	1.50 ± 2.12	4.5 ± 4
PA6.6	1	328 ± 141	402 ± 94	1.66 ± 2.08	9 ± 5
	3	33 ± 44	268 ± 45	2 ± 2.84	3.5 ± 5
PP	1	368 ± 213	213 ± 145	7.33 ± 7.09	12.33 ± 8.50
	3	21 ± 6	112 ± 50	1 ± 0.71	1

paraffin finishing also hinders the identification of microplastic fibers. The contamination from organic matter, such as cellulose and protein matter, is the largest source of material release, accounting for up to 98.5% of the microparticles released. This procedure displays the necessity of precise and well-defined experimental methods that are able to capture the physico-chemical nature of the pollutants and discern them from a large quantity of contaminants. Moreover, it could be necessary to develop pre-treatment methods to mitigate this issue and effectively investigate the microplastic emissions strictly related to the textile product. The additive environmental impact of fibrous microplastics was assessed through the life cycle assessment impact category "*Physical effects on biota*". The potential effects on the marine compartment at midpoint and endpoint level were estimated on average at 5.43e-02 PAF m^3 day/kg_{sample} and 1.48e-06 PDF m^2 year/kg_{sample} for PA6.6, 5.05e-02 PAF m^3 day/kg_{sample} and 1.37e-06 PDF m^2 year/kg_{sample} for PA6, 9.72e-01 PAF m^3 day/kg_{sample} and 2.65e-05 PDF m^2 year/kg_{sample} for PET, and 1.37e-02 PAF m^3 day/kg_{sample} and 3.72e-07 PDF m^2 year/kg_{sample} for PP, respectively. Despite the similarities in microplastic filament emissions, PET fibers display the largest additive burden due to their low degradation rate and persistence in the environment.

Conclusions

Including the effects of microplastics in the LCA methodology is of the utmost importance for a holistic and correct description of the environmental burden of plastic products. A shared consensus on the microplastic leakage mapping along the plastic value chain is still missing, and experimental procedures need further refinement to provide a science-based background in standardization efforts. In the textile context, the presence of contaminants is shown to be responsible for a significant material release and to effectively hinder the detection of microplastic emissions. In this work, a harmonized method to simulate a domestic washing process, discretely characterize the emitted material mass, and provide relevant information such as size and shape for a life cycle assessment description was presented. Further pre-treatment procedures could be needed to ensure the replicability of the results. This methodological approach could be expanded to further life cycle stages to holistically describe the additive environmental burdens of both micro and macro plastics. In the investigated samples, fibrous microplastic emissions were similar across fabrics and the number of washes, reinforcing the role of washing and manufacturing conditions in the discharge of microplastic filaments. Polyethylene terephthalate microfilaments display the largest environmental impact through the novel impact category "*Physical effects on biota*", due to their long persistence in the environment. Despite the recent efforts, current environmental models describing the fate of microplastics in the ecosystem are still incomplete and could lead to inaccurate conclusions. In the future, a multi-compartmental and dynamic description, including accumulation and transport mechanisms, could provide a better ground for comparison of plastic products. Additionally, the ecotoxicity effects of additives and plasticizers, the availability of microplastics as collectors and carriers of pollutants, and fragmentation mechanisms are still poorly described and need further investigation.

References

Browne MA, Crump P, Niven SJ, Teuten E, Tonkin A, Galloway T, Thompson R (2011) Accumulation of microplastics on shorelines worldwide: sources and sinks. Environ Sci Technol 45(21):9175–9179

Corella-Puertas E, Hajjar C, Lavoie J, Boulay AM (2023) MarILCA characterization factors for microplastic impacts in life cycle assessment: physical effects on biota from emissions to aquatic environments. J Clean Prod 418:138197

Dalla Fontana G, Mossotti R, Montarsolo A (2021) Influence of sewing on microplastic release from textiles during washing. Water Air Soil Pollut 232(2):50

Käppler A, Windrich F, Löder MGJ, Malanin M, Fischer D, Labrenz M, Eichhorn KJ, Voit B (2015) Identification of microplastics by FTIR and Raman microscopy: a novel silicon filter substrate opens the important spectral range below 1300 cm-1 for FTIR transmission measurements. Anal Bioanal Chem 407(22):6791–6801

Quantis EA (2020) Plastic leak project: methodological guidelines v1.3

Saadi N, Lavoie J, Fantke P, Redondo-Hasselerharm P, Boulay AM (2025) Including impacts of microplastics in marine water and sediments in life cycle assessment. J Clean Prod 520:146037

Schwarz AE, Lensen SMC, Langeveld E, Parker LA, Urbanus JH (2023) Plastics in the global environment assessed through material flow analysis, degradation and environmental transportation. Sci Total Environ 875:162644

Chapter 29
Functionalization of Natural Polymers with Vanillic Acid Derived from PET Waste

Valentina Marturano, Sarai Agustin-Salazar, Giovanni Dal Poggetto, Giovanna Lo Vecchio, and Pierfrancesco Cerruti

Abstract Plastic pollution in aquatic ecosystems has become a major environmental concern, with polyethylene terephthalate (PET) among the most persistent polymers. This study explores the valorization of vanillic acid (VA)obtained from the bioconversion of PET microplastics collected from Lake Maggiore, Italy. VA, a phenolic compound with antioxidant and antimicrobial properties, was grafted onto chitosan to enhance its stability and functionality. Three grafting methods—free radical-mediated,microwave-assisted, and EDC/NHS coupling—were compared. FTIR analysis confirmed successful grafting, while UV-Vis quantification revealed the highest yield (80%) for the EDC/NHS method. The results demonstrate that EDC/NHS coupling provides an efficient, biocompatible route for producing chitosan-VA conjugates, supporting sustainable strategies for PET upcycling and polymer functionalization.

Introduction

Plastic is one of the most widely used materials to produce everyday objects. However, its widespread use has led to an increasing accumulation of plastic waste in the environment, particularly in aquatic ecosystems, with the consequent formation of compounds deriving from its degradation (Hiraga 2019). Among plastics,

---

V. Marturano (✉) · S. Agustin-Salazar · G. Dal Poggetto · P. Cerruti
Institute of Polymers, Composites and Biomaterials, National Research Council (IPCB-CNR), Pozzuoli, NA, Italy
e-mail: valentina.marturano@cnr.it

V. Marturano · P. Cerruti
Institute of Polymers, Composites and Biomaterials UOS Lecco, National Research Council (IPCB-CNR), Lecco, MI, Italy

G. Lo Vecchio
Department of Biomedical Sciences, University of Messina, Messina, ME, Italy

G. Gentile et al. (eds.), *Proceedings of the 4th International Conference on Microplastic Pollution in the Mediterranean Sea*, Springer Water,
https://doi.org/10.1007/978-3-032-23847-4_29

PET is one of the most widespread polyesters, but only 21% of it is recycled, as the recycled material has a lower quality and a higher cost than virgin PET (George 2014). For example, in Lake Maggiore, the density of microplastics increased from 39,000 MP/km^2 in 2016 to 100,000 MP/km^2 in 2018 (Galafassi 2021). The presence of microplastics in aquatic environments has direct implications on the food chain.

To address the environmental impact of PET, physical and chemical recycling processes have been developed. Among them, organic upcycling represents a promising technology to overcome the economic limitations of conventional recycling methods. This process involves the degradation of PET into simple molecules, such as terephthalic acid and ethylene glycol, which can be further converted into high-value compounds, including adipic acid, catechol, glycolic acid, and vanillic acid (VA) (Rosini 2025).

Vanillic acid is a natural phenolic compound known for its antioxidant, antimicrobial, and anti-inflammatory properties. However, its application is limited by its low water solubility and instability under adverse environmental conditions. It can be, however, used to functionalize natural polymers.

In this study, VA obtained from the bioconversion of microplastics collected from Lake Maggiore was employed in the functionalization of natural polymers, such as chitosan. The study focuses on the comparison of three grafting methods: free radical-induced grafting (Curcio 2015), microwave, and EDC/NHS mediated grafting (Xia 2025).

Materials and Methods

The synthesis of chitosan-vanillic acid conjugates (Ch-g-Va) was performed using, as said, three grafting methods.

- Free Radical mediated grafting

  0.5 g of chitosan were dissolved in 50 mL of 2% (v/v) aqueous acetic acid solution. Then, 1 mL of 1 M H_2O_2 containing 0.054 g of ascorbic acid was added dropwise to the chitosan solutions, under stirring for 30 min. 0.5 g of vanillic acid dissolved in 2.5 mL of absolute ethanol was then added to the solution. The reaction was carried out for 24 h at 25 °C with constant stirring. The reaction mixture was dialyzed with distilled water with a 14,000 Da molecular weight cutoff membrane for 48 h to remove unreacted phenolic acids
- Microwave-assisted grafting

  0.5 g of alginate was dissolved in 50 mL 2% (v/v) aqueous acetic acid solution. Then 0.5 g of vanillic acid were dissolved in 5 mL of ethanol and added dropwise to the chitosan solution under constant stirring for about 30 min. The resulting solution was transferred to a microwave-safe container and microwaved at 300 W for 10 min, maintaining a constant temperature of about 60 °C using Multiwave 5000 (Antoon Paar). After microwave treatment, NaOH solution (0.1 M) was slowly added to neutralize the pH of the solution until a pH of 7.0 was reached.

The neutralization resulted in the precipitation of the Ch-g-Va complex which was filtered and dried under vacuum.

- EDC/NHS couple-mediated grafting

  1.25 g of chitosan was dissolved in an aqueous solution of acetic acid 25 mL, 1%, v/v). 0.937 g of vanillic acid were dissolved in 30 mL of EtOH and 1-ethyl-3-(30-dimethylaminopropylcarbodiimide (EDC, 1.2 mol equivalent to VA) was added to react with VA. N-hydroxy succinimide (NHS, 1.2 mol equivalent to VA) was further added. The reaction was stirred for 1 h in EtOH. The vanillic acid solution was gradually added to the chitosan solution and stirred at room temperature for 48 h. After 48 h, 3 volumes of cold EtOH (300 mL) were added to the solution and kept under stirring for 2 h, washed thoroughly with ethanol several times and centrifuged. (10,000 rpm 30 min) to obtain Ch-g-Va in the two concentrations.

A Fourier transform infrared (FTIR) spectrometer (Perkin-Elmer) was used to obtain infrared spectra of the samples. Spectra were recorded between 4000 and 400 cm^{-1} with a resolution of 4 cm^1 at a total scan rate of 32.

The quantification of vanillic acid in Ch-g-Va was performed via UV-vis spectroscopy at a wavelength of 260 nm. Briefly, calibration curves were constructed at increasing concentrations of vanillic acid and the concentration of vanillic acid in the triplicate sample was determined at the specific wavelength.

Results and Discussion

FTIR Analysis of the grafted samples, reported in Fig. 29.1, confirmed that the grafting was successful. Reported results refer to the Ch-g-VA obtained via EDC/NHS couple mediated grafting, other sample show similar results. The Stretching of the OH group, occurring in chitosan at 3351 cm^{-1}, in the grafted samples Ch-g-Va (Fig. 29.1c) is shifted due to the reaction of the NH_2 groups with chitosan. Furthermore, the band present in the chitosan spectrum at 1652 cm^{-1} relating to the free amino groups is not present in the grafted samples.

The results related to the quantification of vanillic acid, reported in Table 29.1 give an indication of the amount of vanillic acid present in the grafted samples obtained via UV-vis spectroscopy. As shown in Table 29.1 results indicated that the grafting yield was higher in the samples obtained via EDC/NHS couple mediated grafting, reaching as high as 80%, while Free Radical mediated grafting and Microwave assisted grafting resulted in a grafting yield of 56 and 42% respectively.

EDC/NHS coupling chemistry is commonly used for grafting polymer chains onto substrates through the formation of amide bonds. This method is highly selective because it targets carboxylic acids and nucleophilic groups, making it an ideal method for grafting biological molecules, such as proteins, to polymeric surfaces.

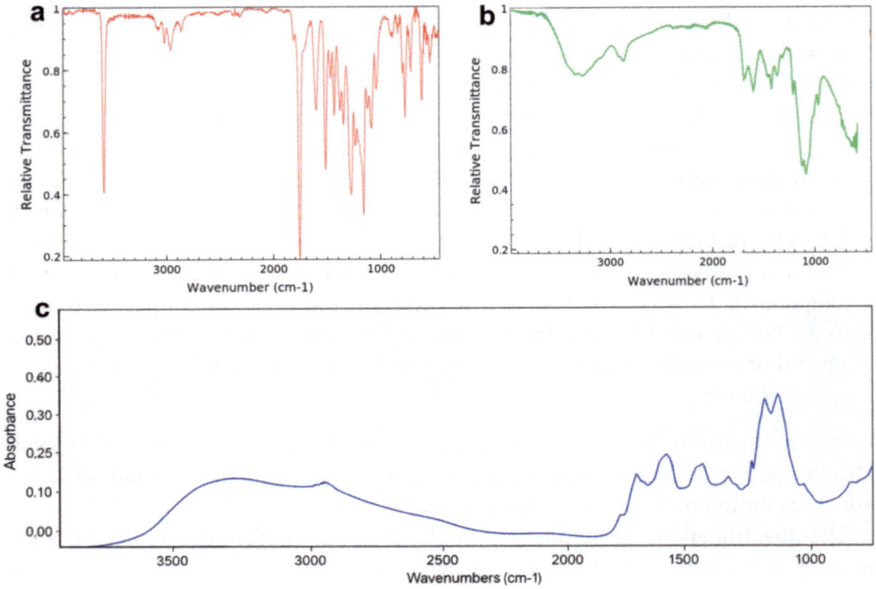

Fig. 29.1 FTIR spectra of vanillic acid (**a**), chitosan (**b**) and Ch-g-VA obtained via EDC/NHS couple mediated grafting (**c**)

Table 29.1 Ch-g-VA grafting yields (%) obtained via UV-vis spectroscopy and compared among three different grafting methods. Concentrations are expressed in mg/L

	Measured VA concentration	Attended VA concentration	Grafting yield (%)
Free radical mediated grafting	0.28	0.5	56
Microwave assisted grafting	0.21	0.5	42
EDC/NHS couple mediated grafting	0.4	0.5	80

Moreover, the reaction occurs under relatively mild conditions (room temperature, aqueous solutions) and avoids the use of potentially hazardous radical initiators, making it a more biocompatible and environmentally friendly option. For these reasons EDC/NHS mediated grafting was selected as the most efficient method for the grafting of VA on chitosan.

Conclusions

This work focused on the valorization of VA resulting from the bacterial and enzymatic conversion of PET collected from Lake Maggiore, Italy.

VA, known for its antioxidant and antimicrobial properties but characterized by high photosensitivity and thermolability. To this end, VA was grafted to Chitosan, a natural polysaccharide. The optimization of the grafting process was performed, comparing various grafting methods.

Acknowledgements This work was supported by the Italian Ministry of University and Research (MUR) under the PRIN 2022 programme (Project No. DCM.PN003.001).

References

Curcio M (2015) Antioxidant polymers: engineered materials as food preservatives and functional foods. Funct Polym Food Sci Technol Biol 209–229

Hiraga K (2019) Biodegradation of waste PET: a sustainable solution for dealing with plastic pollution. EMBO Rep 20(11)

Galafassi S (2021) Microplastic pollution in perch (Perca fluviatilis, Linnaeus 1758) from Italian south-alpine lakes. Environ Pollut 288:117782

George N (2014) Recent developments in the chemical recycling of postconsumer poly(ethylene terephthalate) waste. Ind Eng Chem Res 53(37):14185–14198

Rosini E (2025) Rethinking plastic waste: innovations in enzymatic breakdown of oil-based polyesters and bioplastics. FEBS Open Bio

Xia ZA (2025) Comparative assessment of the antioxidant and antibacterial properties of various phenolic acid-chitosan grafted polymers. J Appl Polym Sci 142(14)

Chapter 30
Preventing Microplastic Formation: CNR Contribution to the EPIC Project's Integrate Response to Marine Macrolitter Accumulation

Silvia Merlino, Cristina De Monte, Giorgio Massaro, Simona Bronco, Enrico Vagnoni, Stefania Solinas, Lucia Ricci, Giovanni Quattrocchi, Marco Bianucci, Fabio Antognarelli, Umberto Granata, Carlo Brandini, Alberto Ribotti, Roberto Sorgente, Giovanni De Falco, and Andrea Cucco

Abstract EPIC (un marE PrIvo di plastiCa) is a strategic project funded by the European Union under the Italy-France Interreg Maritime program. It promotes circular economy actions and innovative solutions to improve resource efficiency and reuse in a transnational context, aiming to generate positive community impacts, foster institutional synergy, and support sector-specific legislation. The project also seeks to advance ecological transition by addressing the issue of macroplastics in marine and coastal environments through shared strategies.

S. Merlino (✉) · M. Bianucci
Institute for Marine Sciences (ISMAR), National Research Council of Italy, Lerici, SP, Italy
e-mail: silvia.merlino@cnr.it

C. De Monte · S. Bronco · L. Ricci · U. Granata
Institute for Chemical and Physical Processes (IPCF), National Research Council of Italy, Pisa, Italy

G. Massaro · G. Quattrocchi · F. Antognarelli · A. Ribotti · R. Sorgente · G. De Falco · A. Cucco
Institute for the Study of Anthropic Impacts and Sustainability in the marine environment (IAS), National Research Council of Italy, Torregrande, OR, Italy

E. Vagnoni · S. Solinas
Institute for Bioeconomy (IBE), National Research Council of Italy, Località Baldinca Li Punti, SS, Italy

C. Brandini
Institute for Marine Sciences (ISMAR), National Research Council of Italy, Florence, Italy

G. Gentile et al. (eds.), *Proceedings of the 4th International Conference on Microplastic Pollution in the Mediterranean Sea*, Springer Water,
https://doi.org/10.1007/978-3-032-23847-4_30

Introduction

EPIC (un marE PrIvo di plastiCa) is a strategic project funded by the European Union under the Italy-France Interreg Maritime program. It promotes circular economy actions and innovative solutions to improve resource efficiency and reuse in a transnational context, aiming to generate positive community impacts, foster institutional synergy, and support sector-specific legislation. The project also seeks to advance ecological transition by addressing the issue of macroplastics in marine and coastal environments through shared strategies. Launched in November 2024, EPIC runs for 48 months and involves 16 partners—10 from Italy and 6 from France. The areas of focus include the Tyrrhenian Sea, NW Sardinia, Corsica, and SE France (Var department).

Materials and Methods

To achieve the project's objectives, it's essential to first understand the main sources of macroplastics entering the sea in the target regions, as well as the processes behind their transport, dispersion, and accumulation. Additionally, assessing the degradation state of plastic objects recovered from the marine environment is crucial for planning recycling strategies and selecting appropriate technologies to support the project's circular economy goals. Two key objectives of the EPIC project are: (1) studying the dispersion and recovery of plastic waste, especially near river mouths; (2) analyzing collected marine debris to determine recycling options and methods. To meet these goals, the 16 project partners have agreed on strategic actions and methodologies. Among these, we report those applied by the CNR partner in its designated areas, i.e. NW Sardinia and the Tyrrhenian Sea, both for the coastal and marine areas (shallow and deep). Especially for the Tyrrhenian Sea area, priority will be given to those areas located at the mouths of rivers, thus also considering the river stretches close to the mouths. A set of drifters, i.e., floating devices that can be tracked via GPS, will be used to identify transport processes of plastic waste at sea. Besides, oceanographic numerical modelling will provide high-resolution information on the transport processes at sea, that could determine accumulation of litter along the Sardinian and Ligurian/Tuscan coasts. CNR team will adopt SHYFEM (Umgiesser et al. 2004; Cucco et al. 2023), a finite element ocean model that solves the primitive equations on unstructured computational grids. Combined with modeling, the drifters will help identify seasonal routes and accumulation areas for floating marine debris, especially macroplastics. Additional analysis using **COPERNICUS** oceanographic data will help identify surface accumulation zones. Field measurements and drone or geophysical surveys, in collaboration with RAS, will validate model results and detect litter accumulations onshore (Fig. 30.1) and underwater.

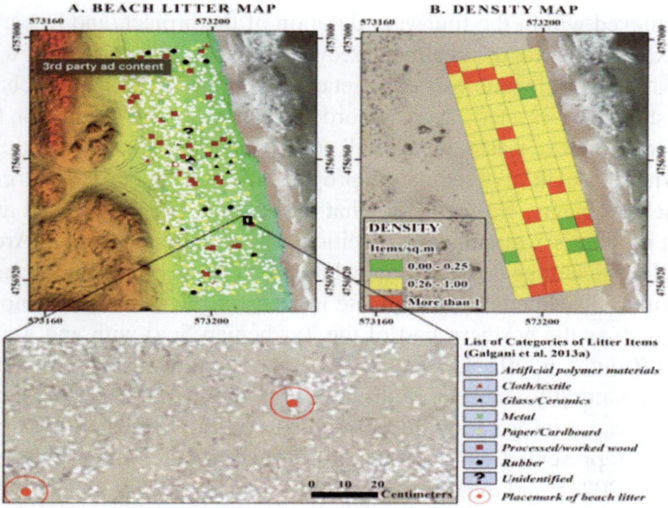

Fig. 30.1 Monitoring waste sources and accumulation areas with aerial drones: after sampling and detection, manual and/or artificial intelligence-supported tools will be used to quantify results, produce maps, and evaluate spatial/temporal distributions

It will develop a protocol to assess chemical and physical aging of fossil- and bio-based plastics, including location, morphology, chemistry, thermal properties, polymer type, degradation, and pollutant absorption. Data obtained by CNR researchers will be shared with project partners working in Corsica and southern France. Specific working groups have been created to integrate data and coordinate technologies (modeling, drifters, satellites, drones) to evaluate their effectiveness in detecting marine plastic waste and optimize recovery and reuse, carefully timing monitoring campaigns and selecting intervention areas, aligned with ARPAT's Marine Strategy, and evaluating available equipment across research groups to ensure consistent settings and comparable data. Finally, a key project goal is to support institutions in defining protocols and recovery actions across river, coastal, and marine environments. Life Cycle Assessment (LCA) studies using circularity indicators and environmental metrics will evaluate long-term impacts of the involved technologies. Results will promote best practices and governance, building on previous projects like Interreg SICOMARplus, GEREMIA, IMPACT, SINAPSI, and AMARE.

Results and Discussion

Since the launch of the project, the 16 partners have begun sharing methodologies and experiences, also taking into account the promotion of the outputs of previous Interreg Maritime projects. From these premises, we expect that EPIC's objectives will be achieved within the four-year duration of the project. and in particular: the existing technologies for the treatment, recycling, reuse, and recovery of waste should be mapped; industrial symbiosis experiences and platforms for the exchange of by-products should be created; a cross-border observatory be set up for the area of cooperation on plastic waste, which will contribute to the implementation of guidelines and the formulation of regulatory proposals. Regarding the CNR's contribution to these achievements, we focus on what has already been addressed and realized in the first projects'six months in a specific botton sea area of Tuscan Archipelagos, Giannutri Island. Unlike floating litter, which often collects in frontal zones (Le et al. 2024), benthic litter tends to accumulate in areas of low circulation, more stagnant currents (such as the northern part of the Tyrrhenian Sea) with and high sediment deposition (Cau et al. 2024). The discovery and identification of submerged accumulations is naturally much more difficult than the identification of surface accumulation areas, so information concerning seabed litter remains relatively scarce: in 2019 they represented 1/4 of all studies on marine litter, microplastics excluded (Canals et al. 2021), and the majority of studies on marine litters have concentrated on floating or beached debris (Galgani et al. 2015; Schwarz et al. 2019). First identified by ARPAT in 2019 in a small, submerged area of Spalmatoio Bay (Giannutri Island), a dense accumulation of plastic waste—mainly bottles—has been systematically studied by CNR since 2021. The research aims to: map the area's morphology, quantify and trace the origin of the waste, estimate its residence time, and assess colonization by benthic species, considering that seabed deposition rates depend on both physical and biological factors, such as depth, currents, wave motion, bottom topography (Madricardo et al. 2020) and the presence of biofouling. Until nowadays, performed activities included bathymetric surveys, seabed clean-ups, analysis of recovered samples and Lagrangian experiments with drifters. With EPIC's launch in late 2024, the site became a key project focus. Using previously collected data, a high-resolution circulation model was implemented, and calibration and analysis of coastal circulation were carried out, with the aim of identifying the causes underlying the frequent and extraordinary accumulation of plastic material on the seabed. A key result is the analysis of water age and trapping dynamics in Spalmatoio Bay. The trapping index, based on tracer residence time, reveals significant differences between surface and depth-averaged layers, highlighting the role of vertical structure in retention (Fig. 30.2). The inner western sector shows the highest trapping values, identifying it as the main accumulation zone. This area is influenced by a cyclonic vortex and semi-enclosed circulation, with weak currents indicating limited exchange with the open sea and hydrodynamic isolation.

The Spalmatoio Bay—particularly its western sector—seems to be an efficient retention area, capable of capturing and holding offshore-introduced materials over prolonged periods. This result may explain why anthropogenic material, including floating material, transported by sea currents to the area around Giannutri Island, remains trapped there, even if it comes from distant areas.

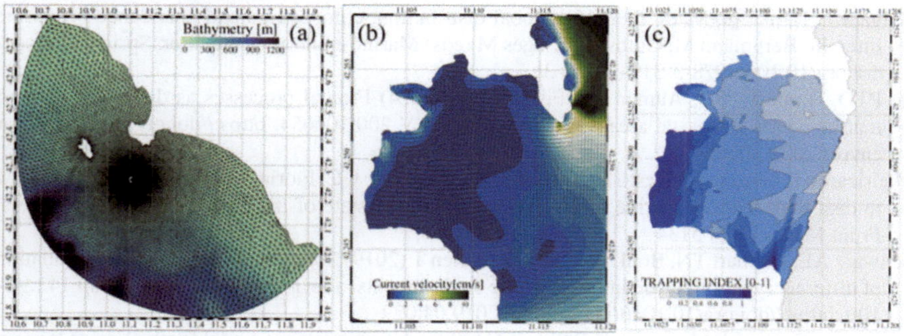

Fig. 30.2 High-resolution numerical model results showing, **a** spalmatoio bay circulation and exchange with the open sea; **b** associated flow structures influencing material trapping; **c** presents the trapping index for the depth-averaged layers, identifying the inner western sector as the main macrolitter accumulation hotspot

Conclusions

The EPIC project tackles environmental and innovation challenges by promoting material circularity. With only 32.5% of plastics recycled today in EU, EPIC goal is to explores traditional and innovative recycling methods to enable diverse reuse options. Key goals include developing multi-territorial strategies, fostering stakeholder collaboration, and promoting dialogue for best practices and regulatory improvements. CNR contributes through four research institutes, using case studies like Giannutri Island to link ocean currents and entrapment maps with plastic accumulation. The multidisciplinary approach combines modeling, biological analysis, surveys, and material studies. Next steps include expanding research to new areas and intensifying plastic aging and degradation analysis to improve recovery methods.

References

Canals M, Pham C, Giorgetti A (2021) The quest for seafloor macrolitter: a critical review of background knowledge, current methods and future prospects. Environ Res Lett 16:023001. https://doi.org/10.1088/1748-9326/abc6d4

Cau A, Sbrana A, Franceschini S, Fiorentino F, Follesa MC, Galgani F, Garofalo G, Gerigny O, Profeta A, Rinelli P, Sbrana M (2024) What, where, and when: spatial-temporal distribution of macro-litter on the seafloor of the western and Central Mediterranean Sea. Environ Pollut 342:123028. https://doi.org/10.1016/j.envpol.2023.123028

Cucco A, Rindi L, Benedetti-Cecchi L, Quattrocchi G, Ribotti A, Ravaglioli C, Cecchi E, Perna M, Brandini C (2023) Assessing the risk of oil spill impacts and potential biodiversity loss for coastal marine environment at the turn of the COVID-19 pandemic event. Sci Total Environ 894:164972. https://doi.org/10.1016/j.scitotenv.2023.164972

Galgani F, Hanke G, Maes T (2015) Global distribution, composition and abundance of marine litter. In: Bergmann M, Gutow L, Klages M (eds) Marine anthropogenic litter. Springer. https://doi.org/10.1007/978-3-319-16510-3t

Le PTD, Hardesty BD, Auman HJ, Fischer AM (2024) Frontal processes as drivers of floating marine debris in coastal areas. Mar Environ Res V 200:106654. https://doi.org/10.1016/j.marenvres.2024.106654

Madricardo F, Ghezzo M, Nesto N, Mc Kiver WJ, Faussone GC, Fiorin R, Moschino V (2020) How to deal with seafloor marine litter: an overview of the state-of-the-art and future perspectives. Front Mar Sci 7:505134

Schwarz AE, Ligthart TN, Boukris E, van Harmelen T (2019) Sources, transport, and accumulation of different types of plastic litter in aquatic environments: a review study. Mar Poll Bull 142:92–100. https://doi.org/10.1016/j.marpolbul.2019.04.029

Umgiesser G, Canu DM, Cucco A, Solidoro C (2004) A finite element model for the Venice lagoon. Development, set up, calibration and validation. J Mar Syst 51(1–4 SPEC. ISS). https://doi.org/10.1016/j.jmarsys.2004.05.009

Chapter 31
Microplastic Contamination in Water and Sediment in Anawilundawa Mangrove Restoration Site, Sri Lanka

Andara Manage Shashini Methsara, Alexandra Mcgoran, Yohan L. Mahagamage, D. S. M. De Silva, Sevvandi Jayakody, and E. M. M. S. Ekanayake

Abstract Microplastics (MPs), originating as primary or secondary pollutants, are pervasive in marine and freshwater systems (Rezania et al. in, Mar Pollut Bull 133:191–208 (2018)). An estimated 10–40 million tonnes are released annually, potentially doubling by 2040 (Thompson et al. in, Science 386:eadl2746 (2024)) posing serious threats to ecosystems due to their persistence, toxicity, and widespread distribution (Frias and Nash in, Mar Pollut Bull 138:145–147 (2019)).

Introduction

Microplastics (MPs), originating as primary or secondary pollutants, are pervasive in marine and freshwater systems (Rezania et al. 2018). An estimated 10-40 million tonnes are released annually, potentially doubling by 2040 (Thompson et al. 2024) posing serious threats to ecosystems due to their persistence, toxicity, and widespread distribution (Frias and Nash 2019). Mangrove ecosystems, at the land-sea interface, act as natural filters but are increasingly affected by terrestrial and marine pollution,

A. M. S. Methsara · D. S. M. De Silva
Department of Chemistry, Faculty of Science, University of Kelaniya, Dalugama, Kelaniya, Sri Lanka

A. Mcgoran
Centre for Environment, Fisheries and Aquaculture Science (CEFAS), Lowestoft, Suffolk, UK

Y. L. Mahagamage
Centre for Environmental Studies and Sustainable Development (CESSD), The Open University of Sri Lanka, Nawala, Nugegoda, Sri Lanka

S. Jayakody · E. M. M. S. Ekanayake (✉)
Department of Aquaculture and Fisheries, Faculty of Livestock, Fisheries and Nutrition, Wayamba University of Sri Lanka, Makandura, Gonawila (NWP), Sri Lanka
e-mail: smeths.er24009@kln.ac.lk

G. Gentile et al. (eds.), *Proceedings of the 4th International Conference on Microplastic Pollution in the Mediterranean Sea*, Springer Water,
https://doi.org/10.1007/978-3-032-23847-4_31

particularly from agricultural, aquacultural, and industrial runoff (Deng et al. 2021) which can alter physicochemical properties of sediment and water, and undermine restoration efforts (Horton et al. 2017; Hu et al. 2023).

Coastal mangrove ecosystems in Sri Lanka face increasing pressure from anthropogenic activities, including plastic pollution from urban runoff, fishing activities, and coastal development (Jayapala et al. 2024). The Anawilundawa Sanctuary, a key site under the Accelerated Natural Regeneration of Mangroves (ANRM) program (WNPSSL 2021), was previously degraded by shrimp farming and is now being restored. Despite its importance, no baseline study has assessed microplastic contamination at the site, and such studies remain limited globally (Mendes et al. 2023). This study examined the abundance, distribution, and characteristics of MPs in surface water and sediment to inform mitigation and restoration strategies (Alazaiza et al. 2022).

Materials and Methods

The study was conducted at the Anawilundawa ANRM site (7°42′41.0″N, 79°48′06.0″E), Sri Lanka, in 2024. Sampling covered ten representative locations, including natural and artificial canals, and mangrove plots of varying ages: RS2 and RS4 (3 + years), RS5-RS7 (11 months), RS8-RS9 (9 months), and RS3, the main canal for plot 1 (3 + years) (→ Fig. 31.1). Sites were selected to reflect spatial variation and the influence of restoration age on MP distribution. At each site, 30 L of surface water and 400 mL of sediment were collected in triplicate. Samples were pre-sieved using a 200 μm sieve, and particles retained (> 200 μm) were analysed. Organic matter was removed by digesting with 10% KOH and 30% H_2O_2, followed by density separation with saturated NaCl (1.2 g/cm^3, 3 times) and NaI (1.8 g/cm^3, once) (Nava and Leoni 2021). The supernatant was filtered through 0.45 μm Whatman GF/F filters and examined under stereomicroscope (Olympus SZ30). Prior to polymer identification, each suspected plastic fragment was verified using the hot needle test (De Witte et al. 2014). Microplastic particles were visually identified and categorized based on shape (fibres, fragments, films, and pellets) and colour. Particles were manually counted on each filter, and microplastic abundance was expressed as items/mL. μ-FTIR analysis will be included in future work to confirm polymer types and improve identification accuracy (Malla-Pradhan et al. 2022). Kruskal-Wallis test was performed to compare MP concentrations across sites and between water and sediment samples. To minimise contamination, all equipment and containers were rinsed with filtered distilled water, and glassware was used where possible. Solutions were pre-filtered (0.45 μm), samples were covered with foil, and surfaces were regularly cleaned. Laboratory and environmental blanks were included, and cotton lab coats and non-synthetic gloves were worn. Equipment was cleaned between uses, and post-filtration samples were dried and stored under contamination-free conditions.

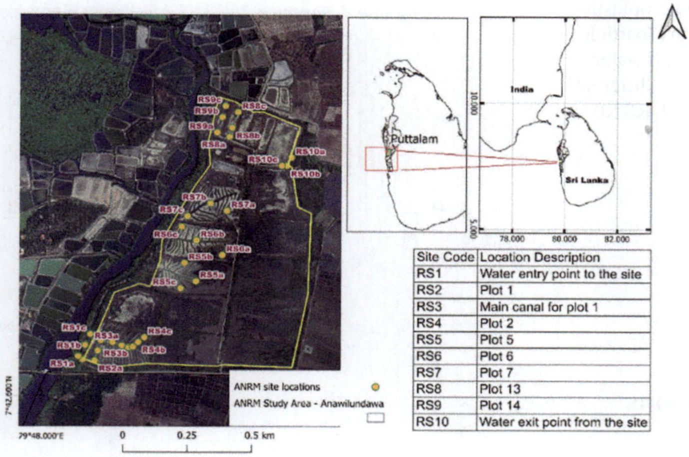

Site Code	Location Description
RS1	Water entry point to the site
RS2	Plot 1
RS3	Main canal for plot 1
RS4	Plot 2
RS5	Plot 5
RS6	Plot 6
RS7	Plot 7
RS8	Plot 13
RS9	Plot 14
RS10	Water exit point from the site

Fig. 31.1 ANRM site study area

Results and Discussion

Microplastic (MP) concentrations in both surface water and sediment samples showed distinct differences between the two environmental matrices. Sediment samples contained substantially higher MP abundance (52.92 ± 24.04 particles/L) compared to water samples (0.41 ± 0.17 particles/L). The total MP count recorded in sediment (529.17 particles/L) was more than 100 times greater than that in water (4.13 particles/L). A Mann-Whitney U test confirmed that sediment concentrations were significantly higher ($U = 0$, $p < 0.0001$), highlighting the sediment-trapping capacity of MPs, especially in low-energy mangrove environments with reduced water flow due to dense root structures.

Kruskal-Wallis tests indicated no significant spatial variation in MPs among the 10 locations for water ($\chi^2 = 8.31$, df = 9, $p = 0.503$) or sediment ($\chi^2 = 11.12$, df = 9, $p = 0.267$), suggesting relatively uniform contamination. The highest sediment MP load occurred at RS3 (89.17 ± 43.47 particles/L) and the highest water contamination at RS7 (0.77 ± 0.58 particles/L), likely due to recent dredging and flow changes (→ Fig. 31.2). Fibres dominated both matrices (81.5% in sediment, 86.2% in water), followed by fragments, indicating inputs from textile and fishing activities. Black and Transparent/White were the most common colours, and MPs < 0.5 mm were the most abundant size class. Control samples showed low but detectable MP contamination in surface water (0.117 ± 0.063 particles/L), consisting solely of fibres. Sediment controls exhibited higher and more variable contamination (10.25 ± 4.48 particles/L), with fibres comprising 78% of MPs.

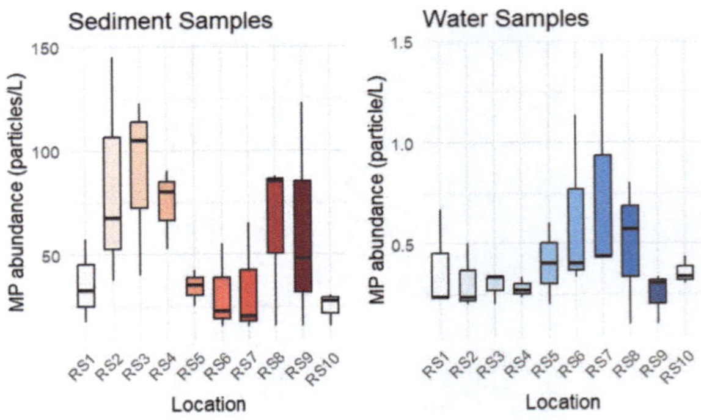

Fig. 31.2 Microplastic concentrations (particles/L) in sediment and water samples across different locations (RS1-RS10)

Conclusions

The present study reveals that the ANRM restoration site acts as an effective sink for sediment MPs, recording a markedly higher mean MP concentration than in water with minimal spatial variation across the site. The results indicate strong anthropogenic influence from point and non-point sources, underscoring the urgent need for integrated waste management, source reduction targeting textile and fishing-related waste, cleanup of hotspots, and continuous monitoring to safeguard these recovering ecosystems from plastic pollution.

Acknowledgements I acknowledge the Ocean Country Partnership Programme (OCPP) and the Association of Commonwealth Universities (ACU) for grant support. I also thank the local authorities and community members for their assistance during fieldwork and site access.

References

Alazaiza MYD, Albahnasawi A, Al-Maskari O, Ali GAM, Eyvaz M, Abujazar MSS, Abu Amr SS, Nassani DE (2022) Microplastic in the environment: identification, occurrence, and mitigation measures. Desalin Water Treat 272:233–247. https://doi.org/10.5004/dwt.2022.28849

De Witte B, Devriese L, Bekaert K, Hoffman S, Vandermeersch G, Cooreman K, Robbens J (2014) Quality assessment of the blue mussel (Mytilus edulis): comparison between commercial and wild types. Mar Pollut Bull 85(1):146–155. https://doi.org/10.1016/j.marpolbul.2014.06.006

Deng H, He J, Feng D, Zhao Y, Sun W, Yu H, Ge C (2021) Microplastics pollution in mangrove ecosystems: a critical review of current knowledge and future directions. Sci Total Environ 753:142041. https://doi.org/10.1016/j.scitotenv.2020.142041

Frias JPGL, Nash R (2019) Microplastics: finding a consensus on the definition. Mar Pollut Bull 138:145–147. https://doi.org/10.1016/j.marpolbul.2018.11.022

Horton AA, Walton A, Spurgeon DJ, Lahive E, Svendsen C (2017) Microplastics in freshwater and terrestrial environments: evaluating the current understanding to identify the knowledge gaps and future research priorities. Sci Total Environ 586:127–141. https://doi.org/10.1016/j.scitotenv.2017.01.190

Hu B, Lin J, He J, Zheng M, Wang J, Nan Y, Wu Y, Guo P (2023) Sediment and interstitial water heavy metals in mangrove restoration wetland and preliminary exploration of microplastics in interstitial water. Catena 221:106764–106764. https://doi.org/10.1016/j.catena.2022.106764

Jayapala HPS, Jayasiri HB, Ranatunga RRMK, Perera IJJUN, Bellanthudawa BKA (2024) Ecological ramifications of marine debris in mangrove ecosystems: estimation of substrate coverage and physical effects of marine debris on mangrove ecosystem in Negombo lagoon, Sri Lanka. Mar Pollut Bull 201:116184. https://doi.org/10.1016/j.marpolbul.2024.116184

Malla-Pradhan R, Suwunwong T, Phoungthong K, Joshi TP, Pradhan BL (2022) Microplastic pollution in urban Lake Phewa, Nepal: the first report on abundance and composition in surface water of lake in different seasons. Environ Sci Pollut Res 29(26):39928–39936. https://doi.org/10.1007/s11356-021-18301-9

Mendes DS, Beasley CR, Silva DNN, Fernandes MEB (2023) Microplastic in mangroves: a worldwide review of contamination in biotic and abiotic matrices. Mar Pollut Bull 195:115552. https://doi.org/10.1016/j.marpolbul.2023.115552

Rezania S, Park J, Md Din MF, Mat Taib S, Talaiekhozani A, Kumar Yadav K, Kamyab H (2018) Microplastics pollution in different aquatic environments and biota: a review of recent studies. Mar Pollut Bull 133:191–208. https://doi.org/10.1016/j.marpolbul.2018.05.022

Thompson RC, Courtene-Jones W, Boucher J, Pahl S, Raubenheimer K, Koelmans AA (2024) Twenty years of microplastic pollution research—what have we learned? Science 386(6720):eadl2746. https://doi.org/10.1126/science.adl2746

WNPSSL (2021) https://www.wnpssl.org/news/anawilundawa.html

Chapter 32
Modeling Nanoplastics in Action: Molecular Structure, Dynamics, Interactions, and Environmental Fate

Giuseppe Milano, Enzo Venezia, Andrea Correa, Alessia Giannattasio, Vincenzo Belgiorno, Carmine Capacchione, and Antonio Buonerba

Abstract Nanoplastics, defined as plastic particles with dimensions below 1 μm, (EFSA 2016) have emerged as a pressing environmental concern due to their potential to accumulation in marine ecosystems and enter food chains. Their small size and high surface area facilitate interactions with living organisms, raising questions about their long-term ecological and health impacts. Studying nanoplastics presents significant challenges: their identification and characterization in complex environmental matrices are extremely difficult, and their behavior at the molecular scale remains poorly understood.

Introduction

Nanoplastics, defined as plastic particles with dimensions below 1 μm, (EFSA 2016) have emerged as a pressing environmental concern due to their potential to accumulation in marine ecosystems and enter food chains. Their small size and high surface area facilitate interactions with living organisms, raising questions about their long-term ecological and health impacts. Studying nanoplastics presents significant challenges: their identification and characterization in complex environmental matrices

G. Milano (✉) · E. Venezia
Department of Chemical, Materials and Production Engineering, University of Naples Federico II, Naples, Italy
e-mail: giuseppe.milano@unina.it

A. Correa
Department of Chemical Science, University of Naples Federico II, Naples, Italy

A. Giannattasio · C. Capacchione · A. Buonerba
Department of Chemistry and Biology "A. Zambelli", University of Salerno, Fisciano, Italy

V. Belgiorno
Department of Civil Engineering, University of Salerno, Fisciano, Italy

G. Gentile et al. (eds.), *Proceedings of the 4th International Conference on Microplastic Pollution in the Mediterranean Sea*, Springer Water,
https://doi.org/10.1007/978-3-032-23847-4_32

are extremely difficult, and their behavior at the molecular scale remains poorly understood.

Experimental characterizations can provide valuable information on the occurrence, size distribution, morphology, and surface chemistry of environmental nanoplastics, but often lack molecular-level resolution. On the other hand, molecular simulations can offer detailed insight into structure, dynamics, and interactions, yet are limited in scale and computational cost when dealing with realistic nanoplastic aggregates in the 10–100 nm range. This scale is particularly relevant, as it bridges molecular structure (including crystalline domains and chemical heterogeneity) with mesoscopic behavior and environmental fate.

To overcome these limitations, we envisage an integrated approach that combines experimental investigation of nanoplastics with multiscale molecular simulations.

Materials and Methods

Nanoplastics Recovery and Characterization

Nanoplastics were recovered from surface seawater following the procedure of Giannattasio et al. (2024). Sampling was conducted between June and December 2021 along the Amalfi Coast-Gulf of Policastro (Tyrrhenian Sea) using glass bottles during night fishing activities. Two liters of seawater were filtered through 0.45 μm nitrocellulose membranes, which were dried, dissolved in acetone, and centrifuged to isolate synthetic polymer debris. The resulting pellets were washed, dried, and analyzed by NMR (Bruker AVANCE 600 MHz). Samples were dissolved in TCE-d_2 at 80 °C, with mesitylene as an internal standard, and polymer molecular weights determined by DOSY NMR using polystyrene standards (Groves 2017).

Molecular Modeling

Models described during the presentation have been obtained using both atomistic and coarse-grained models (CG), suitable for molecular dynamics (MD) simulations. Atomistic simulations have been performed using GROMACS software, while simulations of CG models have been performed using OCCAM code (Zhao et al. 2012). Simulations of CG models have been performed using hybrid particle-field MD, this simulation technique is based on a representation of non-bonded interactions using density fields. This approach can efficiently exploit parallel hardware, allowing the simulation of large systems (Milano et al. 2023).

Results and Discussion

Understanding the behavior of NPs at the molecular level is essential to assess their environmental fate and their interactions with living organisms. Significant challenges hinder the direct investigation of molecular features of NPs. On the experimental side, detecting and characterizing nanoplastics in real environmental samples is extremely complex due to their small size, low concentrations, and the presence of natural organic matter and other particulates. On the simulation side, fully atomistic approaches are computationally unfeasible at the relevant 10–100 nm scale. To bridge this gap, molecular models based on a hybrid-particle field description have been proposed very recently for semi-crystalline PE (Venezia et al. 2025). The aim of these models is to have a CG representation close to an atomic level and able to represent surface chemistries, effects of pH conditions and ions corresponding to typical environmental conditions, but at the same time, having a computational cost suitable to reach the relevant length-scale.

Polyethylene Nanoplastics: Origin, Course, and Fate

A new methodology enables the detection, isolation, and quantification of micro- and nanoplastics in surface seawater (Giannattasio et al. 2024), identifying low-molecular-weight polyethylene (PE), polydimethylsiloxane (PDMS), and traces of polypropylene (PP), typically overlooked by conventional analyses. The procedure involves filtering 2 L seawater samples, dissolving the filter, and recovering particulates for characterization by ^{1}H and ^{13}C NMR, which provides unambiguous polymer identification and insights into microstructure and oxidation. Seawater samples contained oxidized PE with hydroxyl, ketone, and ester groups, unlike the unoxidized PE found in wastewater influent and effluent. DOSY NMR showed the presence of very low-molecular-weight oligomers (1.0–1.5 kDa). The oxidized PE in seawater may thus derive both from bulk PE degradation and from commercial low-molecular-weight PE wax, widely used as an industrial additive. These findings match those of accelerated weathering of LDPE (Menzel et al. 2022), which revealed a stepwise fragmentation process, preferential loss of amorphous regions, aggregation of nanoscale fragments, and progressive oxidation into hydroxyl, ketone, carboxyl, ester, and peroxidic groups, with average molecular weights dropping to ~1 kDa (Fig. 32.1).

Molecular Models of Nanocrystal Aggregates

The mechanism of formation of an aggregate formed by PE nanocrystals is highlighted in Figure 32.2a. According to the mechanism three stage mechanism proposed

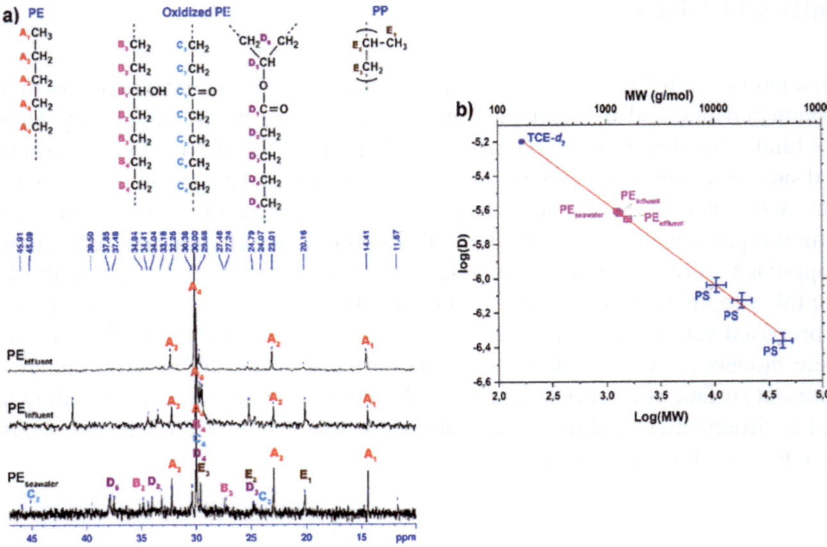

Fig. 32.1 ^{13}C (panel a) and DOSY (panel b) NMR characterization of PE nanoplastics from seawater ($PE_{seawater}$), influent ($PE_{influent}$) and effluent ($PE_{effluent}$) wastewater stream

by Menzel et al. (2022) for semi-crystalline polymers, after surface abrasion (Stage I) and surface breaking (Stage II), in the last stage involving fragment formation and agglomeration, crystalline regions of the polymeric material degrade giving rise to nano-sized crystalline fragments bringing polar functional groups (carboxylic –COOH and hydroxyl groups –OH) on their surface. These small fragments self-assemble in water and form aggregates. According to this mechanism, nanoplastics from semi-crystalline polymers are present in water in the form of aggregates of nano-sized highly crystalline fragments. A recently proposed protocol from our research group, starting from atomistic models, can generate CG molecular models of nano-crystallites corresponding to different stages of aging. (Venezia et al. 2025) In Figure 32.2b, an atomistic model of a crystallite of PE is depicted together with the CG models at different aging times. The colloidal stability obtained for these models is in agreement with different experimental trends regarding their aging and solution conditions as pH, surface charge and counter-ion valence. As shown in Figure 32.2c, these model nano-crystallites can be used for hybrid particle-field MD simulations to obtain aggregates. The obtainment of molecular models of aggregates is the starting point for further studies and comparison with experimental characterization of nanoplastics.

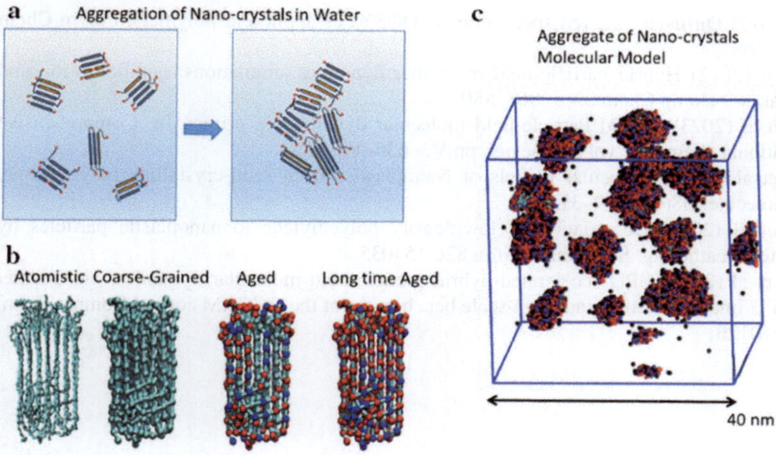

Fig. 32.2 **a** Aggregation of PE nano-crystals formed during the last (third) stage of the three-stage model for the abiotic degradation of a semi-crystalline polymer. **b** Atomistic model (left) of a nano-crystal of PE and corresponding CG models for a pristine non-aged (first from left), aged and long-time aged model. **c** A representative snapshot of aggregate obtained from hybrid particle-field MD simulation of the self-assembly process of 48 PE nano-crystals in water; in the same panel the color code for the CG beads of ions (black), alkyl (magenta), bringing alcohol (red) and carboxyl (blue) groups is reported. Water beads are not shown in the simulation box for clarity

Conclusions

The oral presentation highlighted the role of molecular simulations as a complementary technique to experimental characterization of elusive systems such as nanoplastics. An integrated experimental-simulation approach can overcome current barriers in nanoplastics research. In particular, the difficult characterization of the molecular structures comes from the aggregation of nano-sized fragments. A detailed molecular characterization is pivotal for a better understanding of the environmental fate and biological effects of nanoplastics. Future steps will involve a tighter coupling with experimental activities, target experiments, and, due to the high efficiency of particle-field models on parallel hardware (Esposito et al. 2025), large-scale simulations of the behavior of aggregates.

References

EFSA Panel on Contaminants in the Food Chain EFSA Panel on Contaminants in the Food Chain (2016) Presence of microplastics and Nanoplastics in food, with particular focus on seafood. EFSA J 14

Giannattasio A et al (2024) Micro(nano)plastics from synthetic oligomers persisting in Mediterranean seawater: comprehensive NMR analysis, concerns and origins. Environ Int 190:108839

Groves P (2017) Diffusion ordered spectroscopy (DOSY) as applied to polymers. Polym Chem 8:6700–6708

Zhao Y et al (2012) Hybrid particle-field molecular dynamics simulations: parallelization and benchmarks. J Comp Chem 33(8):868–880

Milano G et al (2023) Hybrid Particle-field molecular dynamics: a primer. In: Comprehensive computational chemistry, vol 3. Elsevier, pp V3–636–V3–659

Venezia E et al (2025) Molecular models of Nanoplastics from semi-crystalline polyethylene. Macromolecules 58(6):3119–3134

Menzel T et al (2022) Degradation of low-density polyethylene to nanoplastic particles by accelerated weathering. Sci Total Environ 826:154035

Esposito R et al (2025) GPU accelerated hybrid particle-field molecular dynamics: multi-node/multi-GPU implementation and large-scale benchmarks of the OCCAM code. J Comput Chem 46(13):e70126

Chapter 33
Advancing Microplastic Mitigation through Life Cycle Assessment: Policy Integration Challenges in the Republic of Serbia within the EU Accession Context

Ljiljana Milošević, Dunja Prokić, Jasna Stepanov, Biljana Panin, Nataša Stojić, Biljana Lončar, and Mira Pucarević

Abstract Over the past decade, microplastics (MPs) have emerged as a major global environmental and public health concern due to their persistence and accumulation across ecosystems (Ali et al. 2024; Ghosh et al. 2023). Addressing this issue is particularly challenging for countries such as the Republic of Serbia, which are still in the process of aligning national legislation with EU environmental standards (Prokić et al. 2025). Life Cycle Assessment (LCA) evaluates environmental impacts throughout the entire life cycle of products or processes, making it a valuable tool for evidence-based policymaking (Bicalho et al. 2017; Fang et al. 2025). However, its application to microplastic pollution is still evolving and requires further investigation (Prokić et al. 2025; Hunt et al.1996; Xayachak et al. 2024). This paper explores key barriers and potential opportunities for integrating LCA into microplasticrelated policy and decision-making in Serbia, within the broader context of global efforts to strengthen environmental governance, particularly in EU-associated countries.

Introduction

Over the past decade, microplastics (MPs) have become a major global environmental and public health concern due to their persistence and accumulation across ecosystems. Inefficient recycling, excessive plastic use, and poorly managed landfills have led to their widespread distribution, causing significant environmental, social, and

L. Milošević (✉) · D. Prokić · J. Stepanov · B. Panin · N. Stojić · M. Pucarević
Faculty of Environmental Protection, Educons University, Sremska Kamenica, Serbia
e-mail: ljiljana.curcic@educons.edu.rs

B. Lončar
Faculty of Technology Novi Sad, University of Novi Sad, Novi Sad, Serbia

G. Gentile et al. (eds.), *Proceedings of the 4th International Conference on Microplastic Pollution in the Mediterranean Sea*, Springer Water,
https://doi.org/10.1007/978-3-032-23847-4_33

economic impacts (Ali et al. 2024; Ghosh et al. 2023). Addressing this issue is particularly challenging for countries like the Republic of Serbia, which are still aligning national legislation with EU environmental standards (Prokić et al. 2025). The main barriers in Serbia include insufficient data availability, weak institutional coordination, and limited research capacity (European Commission 2024; Government of the Republic of Serbia 2010; Government of the Republic of Serbia 2022).

Life Cycle Assessment (LCA) offers a comprehensive, science-based approach to evaluating environmental impacts across the entire life cycle of products or processes, making it a valuable tool for evidence-based policymaking (Bicalho et al. 2017; Fang et al. 2025). However, its application to microplastic pollution is still methodologically evolving and requires further investigation. (Prokić et al. 2025; Hunt et al.1996; Xayachak et al. 2024).

This paper aims to identify the main barriers and emerging opportunities for integrating LCA into decision-making and policy frameworks to address microplastic pollution in Serbia, with broader implications for EU-associated countries seeking to enhance their environmental governance frameworks.

Materials and Methods

This study used a desk review methodology to address its research objectives (Bowron et al. 2019). The scientific literature was systematically reviewed using major academic databases, supplemented by relevant policy documents and national and international reports (Snyder 2019; Prokić et al. 2022).

Results and Discussion

The EU has set challenging goals to reduce microplastic pollution, specifically a planned 30% reduction in microplastic emissions by 2030, as part of the Zero Pollution Action Plan (Munhoz et al. 2023; European Commission 2025 Microplastics). The effective mitigation of microplastic pollution depends on the implementation of comprehensive policies and robust regulations governing the entire life cycle of plastics to minimize microplastic emissions (Nikpay et al. 2024). Recent studies have highlighted the role and increasing importance of LCA in EU decision-making and policy-making, suggesting that extending its application to MPs could provide valuable insights for sharping policies aligned with EU framework (Röck et al. 2021; Sala et al. 2021).

In line with these goals, recent developments in LCA at the international level have increasingly addressed the environmental impacts of microplastics, focusing predominantly on aquatic ecosystems and various polymer types. Notably, several international studies are actively developing characterization factors (CFs) to quantify the physical and ecotoxicological impacts of microplastic emissions, reflecting

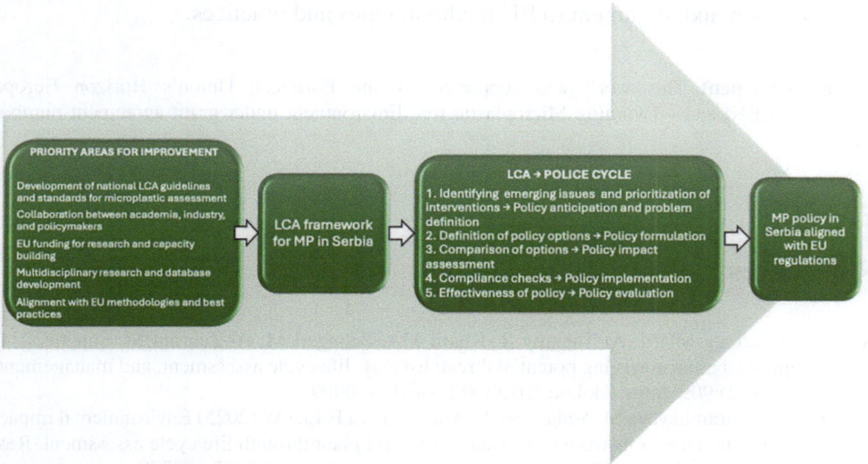

Fig. 33.1 Advancing LCA methodologies for MPs in Serbia to support policy in line with EU regulatory principles (based on Sala et al. 2016)

ongoing efforts to refine LCA methodologies for microplastic assessment (Prokić et al. 2025; Corella-Puertas et al. 2023; Pellengahr et al. 2025; Schwarz et al. 2024; Vázquez-Vázquez et al. 2025; Alibekov et al. 2025). In contrast, studies on LCA in Serbia have primarily focused on plastic waste management, highlighting the need for future research aimed at developing region-specific CFs for microplastics originating from mismanaged plastic in terrestrial environments (Prokić et al. 2025).

Therefore, advancing LCA methodologies for MPs in Serbia could help address this research gap and support the formulation of effective national policies consistent with EU objectives (Fig. 33.1).

Conclusions

To enable LCA to support policymakers in making more transparent and evidence-based decisions regarding microplastic pollution in Serbia, in line with EU Regulation principles, several areas require improvement, including:

- Development of national guidelines and standards for the use of LCA in microplastic assessment.
- Strengthening collaboration between the academic community, industry, and policymakers.
- Utilization of EU funds and international projects to finance research and capacity building.
- Promotion of multidisciplinary research and the development of databases for microplastic.

- Adaptation and alignment to EU methodologies and practices.

Acknowledgement This work was supported by the European Union's Horizon Europe Project GREENLand—Twinning Microplastic-free Environment under grant agreement number 101079267.

References

Ali SS, Alsharbaty MHM, Al-Tohamy R, Khalil MA, Schagerl M, Al-Zahrani M, Sun J (2024) Microplastics as an emerging potential threat: toxicity, life cycle assessment, and management. Toxics 12(12):909. https://doi.org/10.3390/toxics12120909

Alibekov A, Meirambayeva M, Yengsebek S, Aldyngurova F, Lee W (2025) Environmental impact of microplastic emissions from wastewater treatment plant through life cycle assessment. Res. Conserv. Recycl. 962:178378. https://doi.org/10.1016/j.scitotenv.2025.178378

Bicalho T, Sauer I, Rambaud A, Altukhova Y (2017) LCA data quality: a management science perspective. J Clean Prod 156:888–898. https://doi.org/10.1016/j.jclepro.2017.03.229

Bowron RC, Weber JE (2019) The evolving reference desk: a case study. J Acad Librariansh 45(5):102057

Corella-Puertas E, Hajjar C, Lavoie J, Boulay AM (2023) MarILCA characterization factors for microplastic impacts in life cycle assessment: physical effects on biota from emissions to aquatic environments. J Clean Prod 418:138197. https://doi.org/10.1016/j.jclepro.2023.138197

European Commission (2024) Commission staff working document: Serbia 2024 report. Brussels: European Commission. SWD(2024) 695 final. Accompanying the communication from the commission to the European Parliament, the Council, the European economic and social committee and the committee of the regions on EU enlargement policy

European Commission (2025) Microplastics. https://environment.ec.europa.eu/topics/plastics/microplastics_en. Accessed 1 Sep 2025

Fang YX, Wu PZ, Chen S, Li Y, Cui SF, Zhu JX, Cao HZ, Jiang KJ, Zhong L (2025) Prospective LCA towards achieving carbon neutrality goals: framework application and challenges. Environ Impact Assess Rev 111:107733. https://doi.org/10.1016/j.eiar.2024.107733

Ghosh S, Sinha JK, Ghosh S, Vashisth K, Han S, Bhaskar R (2023) Microplastics as an emerging threat to the global environment and human health. Sustainability 15(14):10821. https://www.mdpi.com/2071-1050/15/14/10821

Government of the Republic of Serbia (2010) Decision on the adoption of the national environmental protection program. Official Gazette of the Republic of Serbia, Belgrade No. 353–459/2010-1

Government of the Republic of Serbia (2022) Program for the development of the circular economy in the Republic of Serbia for the period 2022–2024. Official Gazette of the Republic of Serbia, Belgrade

Hunt RG, Franklin WE, Hunt RG (1996) LCA—How it came about. Int J Life Cycle Assess 1:4–7. https://doi.org/10.1007/BF02978624

Munhoz DR, Harkes P, Beriot N, Larreta J, Basurko OC (2023) Microplastics: a review of policies and responses. Microplastics 2:1–26. https://doi.org/10.3390/microplastics2010001

Nikpay M, Toorchi Roodsar S (2024) Crafting a scientific framework to mitigate microplastic impact on ecosystems. Microplastics 3(1):165–183. https://doi.org/10.3390/microplastics3010010

Pellengahr F, Corella-Puertas E, Mattelin V, Saadi N, Bertella F, Boulay AM, van der Meer Y (2025) Modeling marine microplastic emissions in life cycle assessment: characterization factors for biodegradable polymers and their application in a textile case study. Front Toxicol 7:1494220. https://doi.org/10.3389/ftox.2025.149422

Prokić D, Lj Ć, Stepanov J, Stojić N, Pucarević M (2022) The role of circular economy in food waste management in fulfilling the United Nations' sustainable development goals. Acta Univ Sapientiae, Aliment 15:51–66

Prokić D, Stepanov J, Milošević L, Panin B, Stojić N, Pucarević M (2025) Evaluating plastic waste management in EU accession countries: a life cycle perspective from the Republic of Serbia with microplastic implications. Sustainability 17(14):6297. https://doi.org/10.3390/su17146297

Röck M, Baldereschi E, Verellen E, Passer A, Sala S, Allacker K (2021) Environmental modelling of building stocks—an integrated review of life cycle-based assessment models to support EU policy making. Renew Sustain Energy Rev 151:111550. https://doi.org/10.1016/j.rser.2021.111550

Sala S, Reale F, Cristobal-Garcia J, Marelli L, Pant R (2016) Life cycle assessment for the impact assessment of policies: life thinking and assessment in European policies and for evaluating policy options. Publications Office of the European Union. JRC Technical Report, Luxembourg

Sala S, Martino A, Antoine A, Fulvio B (2021) The evolution of life cycle assessment in European policies over three decades. Int J Life Cycle Assess. https://doi.org/10.1007/s11367-021-01893-2

Schwarz AE, Herlaar S, Cohen QM, Quik JTK, Golkaram M, Urbanus JH, van Emmerik THM, Huijbregts MAJ (2024) Microplastic aquatic impacts included in life cycle assessment. Res Conserv Recycl 209:107787. https://doi.org/10.1016/j.resconrec.2024.107787

Snyder H (2019) Literature review as a research methodology: an overview and guidelines. J Bus Res 104:333–339

Vázquez-Vázquez B, Lazzari M, Hospido A (2025) Terrestrial characterization factors for bio- and fossil-based plastics: microplastics ingestion and additives release. Waste Manag. 196:106–114. https://doi.org/10.1016/j.wasman.2025.02.008

Xayachak T, Haque N, Lau D, Pramanik BK (2024) The missing link: a systematic review of microplastics and its neglected role in life-cycle assessment. Sci Total Environ 954:176513. https://doi.org/10.1016/j.scitotenv.2024.176513

Chapter 34
Greenplasma®: a Mobile Waste-to-Energy System for Marine Plastic Recovery and Valorisation

Tommaso Mosca, Luca Giammichele, Alessandro Nardi, Lucia Pittura, Stefania Gorbi, Carlo Ferraro, Manuel Lai, and Francesco Regoli

Abstract Marine plastic pollution is a critical global challenge, with millions of tons of waste entering the oceans each year and very limited options for managing the highly degraded materials found at sea. Despite growing public and policy attention, and the rise of circular economy frameworks, end of life pathways for marine plastics remain largely restricted to landfilling or incineration, generating environmental,

Tommaso Mosca and Luca Giammichele: Co-first authorship.

T. Mosca · L. Giammichele · A. Nardi · L. Pittura · S. Gorbi · F. Regoli (✉)
Department of Life and Environmental Sciences, Marche Polytechnic University, Ancona, Italy
e-mail: f.regoli@univpm.it

T. Mosca
e-mail: tommaso.mosca@iusspavia.it

L. Giammichele
e-mail: l.giammichele@univpm.it

A. Nardi
e-mail: a.nardi@univpm.it

L. Pittura
e-mail: l.pittura@univpm.it

S. Gorbi
e-mail: s.gorbi@univpm.it

T. Mosca
University School for Advanced Studies IUSS, Pavia, Italy

L. Giammichele
Department of Industrial Engineering and Mathematical Sciences, Marche Polytechnic University, Ancona, Italy

A. Nardi · L. Pittura · S. Gorbi · F. Regoli
National Biodiversity Future Center, Palermo, Italy

C. Ferraro · M. Lai
IRIS S.R.L., Orbassano, Italy
e-mail: carlo.ferraro@irssrl.eu

G. Gentile et al. (eds.), *Proceedings of the 4th International Conference on Microplastic Pollution in the Mediterranean Sea*, Springer Water,
https://doi.org/10.1007/978-3-032-23847-4_34

logistical, and socio economic drawbacks. In addition to effective policy-making, addressing this gap requires innovative solutions capable of treating non-recyclable waste while avoiding the negative impacts associated with current disposal practices. This study presents Greenplasma®, a mobile waste to energy (WtE) system designed to convert non-recyclable marine plastics into a hydrogen rich syngas through high-temperature pyrolysis. Fully integrated into a van, the system enables on site treatment along coastlines and in remote areas, reducing transport needs and supporting emerging governance frameworks such as Italy's SalvaMare law. This pilot application on plastics collected at sea confirmed the production of a syngas suitable for electricity generation. Overall, Greenplasma® appears to offer a scalable technological pathway that links marine litter recovery, circularity principles, and coastal resilience, contributing to the development of integrated strategies for sustainable marine ecosystem conservation.

Introduction

Plastic is a valuable material, yet its end-of-life management remains highly problematic, particularly due to the environmental implications of abandoned waste. Each year, millions of tons of plastic enter the oceans and disperse across marine environments (Jambeck 2015; OECD 2022), becoming a major global concern that affects nearly all forms of marine life, ecosystems functioning and biodiversity (Costa 2025; UNEP 2021). Also, sectors relying on healthy oceans—such as fisheries and coastal tourism—are increasingly vulnerable (Mcilgorm 2023; Karasik 2022). Despite growing awareness, projections indicate that up to 53 million tons of plastic could enter the environment annually by 2030, while the capacity to recover or manage this waste remains very limited (Cowan 2021).

Governance efforts increasingly highlight the need for cross-sectoral strategies to reduce the amount of plastic indiscriminately entering the environment, including global policy development, circular economy approaches, and enhanced corporate responsibility (Zhou 2024). Also, clean-up strategies and technological innovation are considered essential components for achieving environmental objectives under the SDGs and European directives (European Commission, 2008; Cunha 2023). Within this context, initiatives such as Italy's "SalvaMare" law (n. 60/2022) support a shift towards circular approaches by actively involving fisher-men in marine litter collection. Nevertheless, plastics recovered at sea don't have virtually any recycling options due to extensive weathering and material degradation and are consequently more likely to end up in landfills or incineration rather than being effectively recycled. These disposal options pose several drawbacks, as chemicals leaching, limited landfill space, and possible public health risks due to pest proliferation (Khoaele 2023).

M. Lai
e-mail: manuel.lai@irissrl.eu

Additional logistical challenges, costs, and greenhouse gas emissions arise when transporting these materials from collection sites to landfills.

Building on previous experience, we present the mobile waste-to-energy (WtE) system Greenplasma®, designed to address this challenge. The prototype converts non-recyclable marine plastics into a hydrogen-rich syngas through high-temperature pyro-gasification, subsequently used for generating electricity. Green-plasma® has been fully integrated into a van, enabling a mobile, on-site waste-to-energy conversion along coastlines and in remote areas. By transforming non-recyclable plastic into usable energy directly at the collection site, the system aligns with circularity principles and offers a scalable framework for coastal resilience, marine pollution mitigation, and sustainable waste management.

Materials and methods

Greenplasma® system is presented in Fig. 34.1. During high-temperature pyrolysis (> 700 °C), plastics polymers are broken into smaller molecules, turning into a variety of products, whose composition depends on the operating conditions, heating rate, temperature, duration of the treatment, and presence of catalysts. The main products of plastic pyrolysis are syngas, pyrolytic oil and char (Dai 2022). In a high-temperature process, as the one adopted in the Greenplasma® plant, the main product is the syngas. The core of the plant is the reactor (Fig. 34.1), a 30 L steel cylinder equipped with electric resistances, for a total power of 4 kW, that provides thermal power necessary to reach and maintain operational temperature between 700 and 900 °C. The syngas produced in the reactor is then passed through: a mechanical cyclone for the separation of the heaviest particles and floating ash, six filters/demisters that absorb CO+, H+S e HCl and reduce the humidity of the syngas by condensation of liquids, and finally, a condenser which decreases the temperature of the syngas in order to intercept the liquid particles still present. The syngas movement and vacuum in the plant are ensured by a blower, equipped with a HEPA (High Efficiency Particulate Air) filter to collect any impurities still present in the syngas that could clog it. Finally, the syngas can be directed to an electric generator with rated power of 5.6 kW or to the flare (during the transient start-up phase of the process or in emergency cases). The system is equipped with an automated monitoring system comprising K-type thermocouples, differential pressure transducers, and a power meter, enabling continuous control of temperature, pressure, blower load, and energy consumption. Syngas composition (CO, CO_2, CH_4, O_2, H_2, H_2S) was measured in real time using an MRU SWG100 gas analyzer, which also calculated the lower heating value (LHV), with a maximum uncertainty of 2% of the reading. The Greenplasma® system, capable of processing up to 10 kg/h of plastic waste, includes two additional units: the shredder and the energy storage system. The shredder is used to reduce plastic waste into pieces smaller than 5 cm before their introduction in the system through a conveyor belt, ensuring a better heat distribution inside the reactor and thus facilitating the decomposition of the materials during

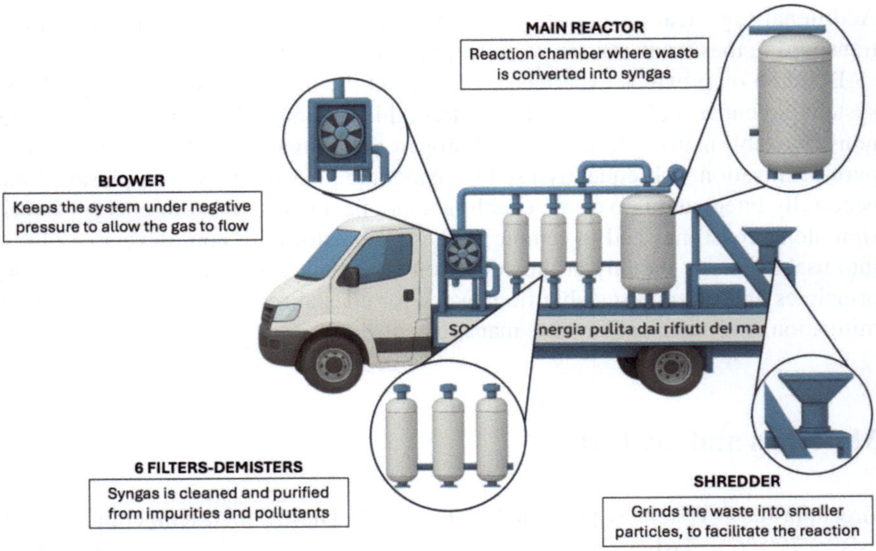

Fig. 34.1 Representation of Greenplasma® mobile plant

pyrolysis. The energy storage system is composed of several components: batteries (5 modules with a total energy equal to 20 kWh), 3 inverters (5 kW each), battery management system (BMS), and a monitoring and data acquisition system which allows to measure both the electrical power used to charge the batteries or supplied by them. The Greenplasma® can be supplied by the grid or by the batteries of the storage system. The latter can be very useful when the Greenplasma® operates in remote zones where it is difficult to have electric energy from the grid. The electric energy produced by the generator can be stored in the batteries for later use or directly employed by an end user for other purposes, such as the charge of electric vehicles.

With the aim of evaluating the composition of the syngas and its LHV, the following pilot test was conducted: 8 kg of polyamide fishing nets collected at sea (previously shredded) were introduced in the reactor, after this had reached an operational temperature of 750 °C. The syngas composition along with its LHV were analyzed for the duration of the whole test (2 h).

Results and discussions

The results of the pilot test, are presented in Fig. 34.2 and confirmed the capability of Greenplasma® to transform 8 kg of non-recyclable plastic waste collected at sea into 5 kg of syngas with a meaningful LHV, the main feature in energy generation. H_2 relative abundance in the syngas reached the peak (~ 35%) after the process reached stability (~ 30 min after the beginning of the high-temperature pyrolysis)

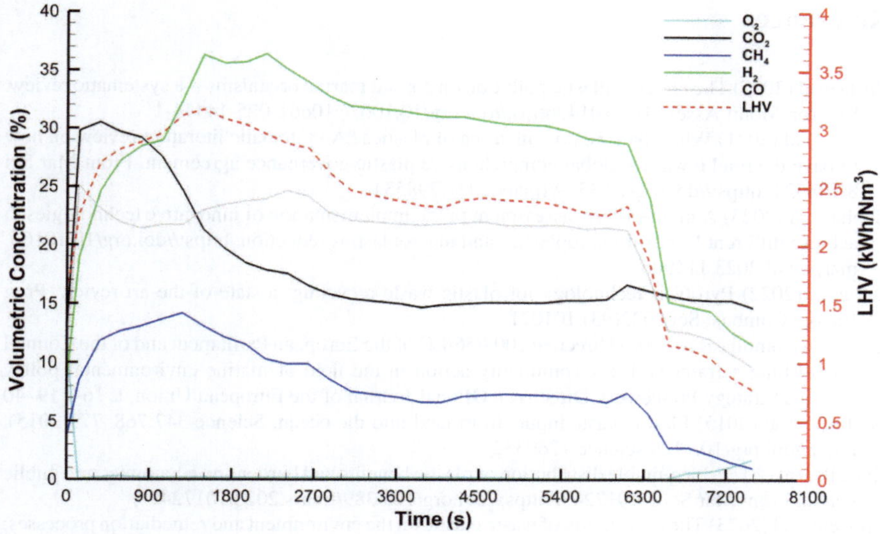

Fig. 34.2 Relative concentrations of different gases in the syngas and lower heating value (LHV, red dashed line) during 2 h pilot test

and remained stable for about 1 h and a half. CH_4 showed a similar behavior, but with lower values. Throughout the test, LHV followed H_2 and CH_4 profile, and was assessed at a value of 2.5–3 kWh/Nm^3, higher than traditional gasification standards (1.5–1.6 kWh/Nm^3). Also, 1.5 kg of char and 1.5 kg of pyrolytic oil were produced. Oxygen profile was consistent with small, unavoidable air intrusions during plastic loading operations at the beginning of the process, and after that O_2 levels rapidly decreased to near-zero values. After reaching theses peaks, the evolution of different gases started to decline, according to the consumption and depletion of plastics loaded into the system.

Conclusions

Overall, the substantial presence of H_2 and CH_4 in the syngas and its meaningful LHV confirmed proper operation of the system and the effectiveness of the pyrolysis process. Moreover, the trial demonstrated the possibility to convert non-recyclable plastic waste collected at sea into a syngas suitable for electricity generation, thereby validating the operational robustness and energy recovery potential of the Greenplasma® system. Lastly, the results indicate that Greenplasma® represents a promising research and development opportunity to support sustainable and circular approaches for marine ecosystem conservation.

References

Costa et al (2025) The effect of plastic pollution on coastal marine organisms—a systematic review. Environ Monit Assess 197:1014. https://doi.org/10.1007/s10661-025-14444-1

Cowan et al (2021) What shall we do with a sea of plastics? A systematic literature review on how to pave the road toward a global comprehensive plastic governance agreement. Front Mar Sci 8:798534. https://doi.org/10.3389/fmars.2021.798534

Cunha et al (2023) A multi-criteria assessment of the implementation of innovative technologies to achieve different levels of microplastics and macroplastics reduction. https://doi.org/10.1016/j.marpolbul.2023.114906

Dai et al (2022) Pyrolysis technology for plastic waste recycling: a state-of-the-art review. Prog Energy Combust Sci 2022(93):101021

European Commission (2008) Directive 2008/56/EC of the European Parliament and of the Council establishing a framework for community action in the field of marine environmental policy (Marine Strategy Framework Directive). Official Journal of the European Union, L 164, 19–40

Jambeck et al (2015) Plastic waste inputs from land into the ocean. Science 347:768–771(2015). https://doi.org/10.1126/science.1260352

Karasik et al (2022) Inequitable distribution of plastic benefits and burdens on economies and public health. Front Mar Sci 9:1017247. https://doi.org/10.3389/fmars.2022.1017247

Khoaele et al (2023) The devastation of waste plastic on the environment and remediation processes: a critical review. Sustainability 15(6):5233. https://doi.org/10.3390/su15065233

Mcilgorm et al (2023) The costs of environmental degradation from plastic pollution in selected coastal areas in the United Republic of Tanzania: Technical Report (English). World Bank Group, Washington, D.C.

OECD (2022) Global Plastics Outlook: Economic Drivers, Environmental Impacts and Policy Options

UNEP (2021) From Pollution to Solution: A Global Assessment of Marine Litter and Plastic Pollution

Zhou L (2024) The global governance of marine plastic pollution: re-thinking the extended producer responsibility system. Front Mar Sci 11:1363269. https://doi.org/10.3389/fmars.2024.1363269

Chapter 35
Wash-Resistant rGO/PU-Coated Cotton Fabrics for Wearable E-Textiles

Federico Olivieri, Roberto Avolio, Rachele Castaldo, Mariacristina Cocca, Maria Emanuela Errico, Marino Lavorgna, and Gennaro Gentile

Introduction

Textile durability under repeated laundering is a major performance criterion for wearable electronic fabrics, particularly when functional coatings are applied to impart electrical or sensing capabilities. Standard textile wash tests are used to evaluate how treatments and coatings withstand mechanical and chemical stress during domestically relevant washing cycles.

GO coatings are extensively used to impart a wide range of properties to suitable substrates, including barrier performance, electrical conductivity, and antibacterial activity (Guerritore 2022; Alp 2024). In this context, smart textiles functionalized with graphene-based fillers represent a widely explored research field (Faggio 2024).

The development of cotton fabrics coated with reduced graphene oxide (rGO) and protected with a thin polyurethane (PU) overlay to enhance durability and preserve functionality after repeated wash cycles (Olivieri 2023).

The primary washing evaluation in this work investigates how coatings affect the visual integrity and electrical properties of the treated cotton fabric after a series of standardized laundering cycles. Understanding wash durability is essential to assess real-world applicability of wearable e-textiles, where resistance to wash-induced degradation determines both functionality and lifespan.

F. Olivieri (✉) · R. Avolio · R. Castaldo · M. Cocca · M. E. Errico · G. Gentile
Institute for Polymers, Composites and Biomaterials National Research Council of Italy, Pozzuoli, Naples, Italy
e-mail: federico.olivieri@cnr.it

M. Lavorgna
Institute for Polymers, Composites and Biomaterials National Research Council of Italy, Portici, Naples, Italy

G. Gentile et al. (eds.), *Proceedings of the 4th International Conference on Microplastic Pollution in the Mediterranean Sea*, Springer Water,
https://doi.org/10.1007/978-3-032-23847-4_35

Materials and Methods

Plain-weave cotton fabric (230 g/m^2, ~ 1 mm thickness) was used as the substrate for functional coatings. Two types of coated fabrics were prepared: (i) cotton with only reduced graphene oxide (rGO) coating and (ii) cotton coated with rGO followed by a polyurethane (PU) protective layer at varying dry contents (0.5 wt%, 1.0 wt%, and 2.0 wt%). The PU overlayer was designed to enhance wash durability without significantly affecting the hand of the fabric (De Falco et al. 2019; Li et al. 2019). To simulate repeated laundering, fabrics were subjected to controlled washing using a Gyrowash laboratory washer (James H. Heal & Co). Each sample was placed in a steel container along with 10 stainless steel balls to mimic mechanical agitation. A distilled water liquor with liquor ratio 150 mL per gram of fabric was used, consistent with standardized textile wash test protocols. Each washing cycle consisted of a 45 min wash period at a defined temperature. The washing protocol included five cycles at 20 °C (cold wash cycles) and five further cycles at 40 °C (hot wash cycles). Figure 35.1 shows a schematic representation of the entire process.

Between subsequent washes, samples were dried at room temperature to emulate typical wearables maintenance. This protocol allows evaluation of washing impact across a representative temperature range for domestic laundering (De Falco et al. 2018).

After laundering, the electrical surface resistivity measurement, evaluated by a Keithley Electrometer/High resistance meter, Model 6517A, with a two-point probe in alternated current, applying 100 V voltage, and reported in ohms per square (Ω/sq), was carried out to evaluate the resistance of the conductive coating. Furthermore, a scanning electron microscopy (SEM) imaging of the cotton fibers before and after washing was performed to observe morphological integrity and coating continuity. A FEI Quanta 200 FEG SEM in high vacuum mode was used and samples were observed at 10 kV acceleration voltage using a secondary electron detector. Before

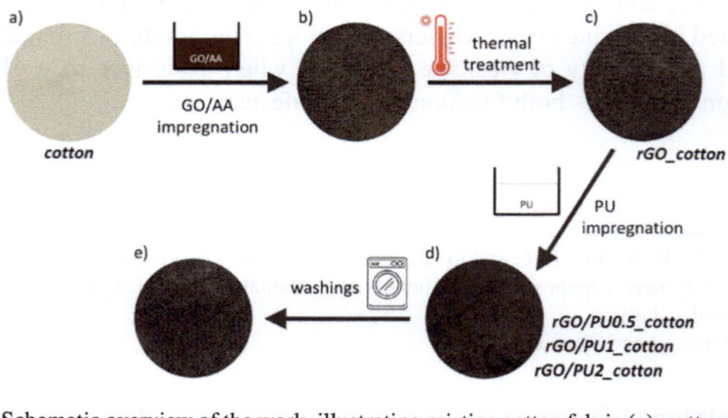

Fig. 35.1 Schematic overview of the work, illustrating pristine cotton fabric (a), cotton fabric after GO/AA impregnation (b), rGO_cotton (c), and rGO/PU0.5_cotton, rGO/PU1_cotton, rGO/PU2_cotton before (d) and after ten washing cycles (e)

SEM observations, samples were mounted onto SEM stubs by means of carbon adhesive disks and sputter coated with a 5–10 nm thick Au–Pd layer.

These analyses provide quantitative and qualitative measures of how well the coatings resist wash-induced alteration or removal.

Results and Discussion

After 10 sequential wash cycles (5 at 20 °C and 5 at 40 °C), neither the rGO-only nor the rGO/PU coated fabrics exhibited significant changes in overall visual appearance. This suggests that the dip-coating and subsequent thermal/chemical curing procedures produced sufficiently adherent coatings on the cotton substrate.

Surface resistivity of coated fabrics (Fig. 35.2) was monitored after each fabrication step and after 5 washes at each temperature condition. All coated samples maintained their conductive properties within the same order of magnitude throughout the washing protocol, indicating limited degradation of the functional rGO network. The fabrics with higher PU content tended to show the lowest resistivity variation over repeated washes, highlighting the role of the PU protective layer in preserving electrical function.

SEM imaging (Fig. 35.3) after the hot wash cycles revealed more pronounced detachment or disruption of rGO sheets in the rGO-only fabrics. In contrast, rGO/PU coated fabrics, particularly those with higher PU content, exhibited fewer signs

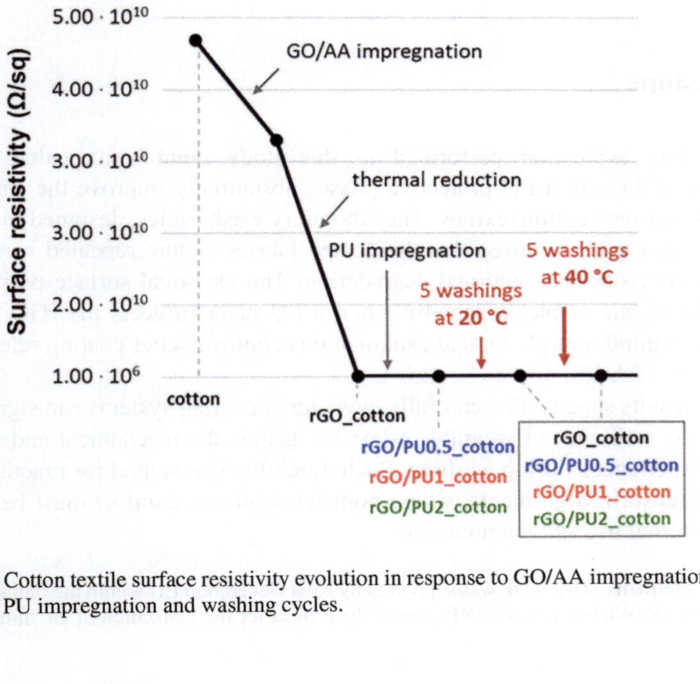

Fig. 35.2 Cotton textile surface resistivity evolution in response to GO/AA impregnation, thermal reduction, PU impregnation and washing cycles.

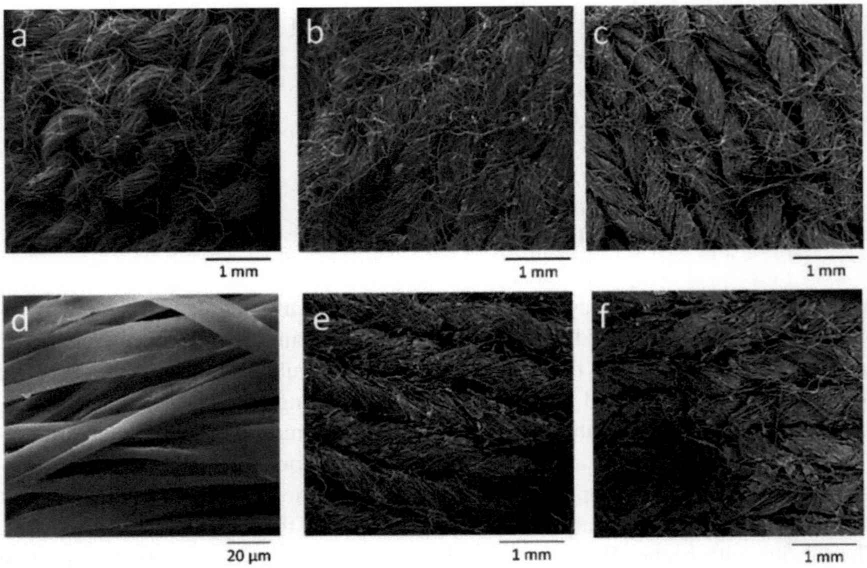

Fig. 35.3 SEM images of cotton (**a, d**), rGO_cotton (**b, e**) and rGO/PU2_cotton (**c, f**) as prepared (**a–c**) and after 10 washings (**e, f**)

of coating detachment. The PU phase acted as a mechanical binder around the fibers, enhancing structural continuity during laundering stress.

Conclusions

The washing evaluation performed in this study demonstrates that coatings combining rGO with a PU protective phase substantially improve the wash durability of functional cotton textiles. The laboratory wash cycles, designed to simulate domestic laundering, showed that the coated fabrics endure repeated wash cycles with limited visual or functional degradation. The electrical surface properties of the textile remain stable, especially when a PU layer protects the rGO network. With this in mind, morphological examinations confirm better coating retention for PU-protected fabrics.

These results suggest that carefully engineered coating systems can significantly enhance the resilience of wearable e-textiles against the mechanical and chemical stresses experienced during washing. Such durability is essential for practical application of sensorized garments, where both function and comfort must be retained through normal use and maintenance.

Acknowledgements This work was supported by Next Generation EU within the Italian National Recovery and Resilience Plan (NRRP), under the project for the realization of an "Infrastructure for Energy Transition and Circular Economy @ EuroNanoLab" (iENTRANCE@ENL), funded by MUR (call Research Infrastructures D.D. n. 3264-IR/2021, CUP: B33C22000710006), and the Extended Partnership PE00000004 "Made in Italy Circolare e Sostenibile" (MICS) project, CUP B53C22004100001, funded by the European Union.

References

Alp E, Olivieri F, Aulitto M et al (2024) The effect of ZnO nanoparticles morphology on the barrier and antibacterial properties of hybrid ZnO/graphene oxide/montmorillonite coatings for flexible packaging. SurfS Interfaces 55:105307

De Falco F, Gentile G, Avolio R et al (2018) Pectin based finishing to mitigate the impact of microplastics released by polyamide fabrics. Carbohydr Polym 198:175–180. https://doi.org/10.1016/j.carbpol.2018.06.062

De Falco F, Cocca M, Guarino V et al (2019) Novel finishing treatments of polyamide fabrics by electrofluidodynamic process to reduce microplastic release during washings. Polym Degrad Stab 165:110–116. https://doi.org/10.1016/j.polymdegradstab.2019.05.001

Faggio N, Olivieri F, Bonadies I et al (2024) Bio-based epoxy resin/carbon nanotube coatings applied on cotton fabrics for smart wearable systems. J Colloid Interface Sci 670:337–347

Guerritore M, Olivieri F, Castaldo R et al (2022) Recyclable-by-design mono-material flexible packaging with high barrier properties realized through graphene hybrid coatings. Resour Conserv Recycl 179:106126

Li H, Zhang W, Ding Q et al (2019) Facile strategy for fabrication of flexible, breathable, and washable piezoelectric sensors via welding of nanofibers with multiwalled carbon nanotubes (MWCNTs). ACS Appl Mater Interfaces 11:38023–38030. https://doi.org/10.1021/acsami.9b10886

Olivieri F, Rollo G, De Falco F et al (2023) Reduced graphene oxide/polyurethane coatings for wash-durable wearable piezoresistive sensors. Cellulose 30:2667–2686. https://doi.org/10.1007/s10570-023-05042-w

Chapter 36
Integrated Assessment of Airborne Microplastics in Plastic-Processing Facilities

Lorenzo Palumbo, Claudia Campanale, Augusta Piccardi, Marcello De Rossi, Jacopo La Nasa, Leonardo Barlucchi, and Francesca Modugno

Introduction

Airborne microplastics (AMPs) are increasingly recognized as an emerging occupational hazard (Boccia et al. 2024; Zarus et al. 2023) due to their persistence, small size, and potential to penetrate deep into the human respiratory tract (Gasperi et al. 2018). While environmental studies have widely documented the ubiquity of microplastics (Kochanek et al. 2025; Loganathan and Kizhakedathil 2023), the occupational dimension remains comparatively underexplored, despite the fact that plastic-processing industries are likely to represent hotspots of exposure. Operations such as cutting, grinding, shredding, and secondary processing release particles into the air, potentially generating significant concentrations within production halls.

Workers in such environments may thus be chronically exposed to a complex mixture of plastic particles of different sizes and chemical compositions. Despite these risks, standardized and validated methodologies for the quantification and characterization of airborne microplastics in occupational settings remain limited.

The present study addresses this gap by developing and validating an integrated analytical protocol that combines active air sampling, fluorescence microscopy (Kalaronis et al. 2022), and Py-GC-MS (Jiang et al. 2025), aiming to provide both quantitative and qualitative insights into workplace exposure.

L. Palumbo (✉) · C. Campanale · A. Piccardi
Department of Environmental and Health, National Institute of Health (ISS), Rome, Italy
e-mail: lorenzo.palumbo@iss.it

M. De Rossi
Laboratory of Industrial Hygiene, ASL Viterbo, Viterbo, Italy

J. La Nasa · L. Barlucchi · F. Modugno
Department of Chemistry and Industrial Chemistry, University of Pisa, Pisa, Italy

G. Gentile et al. (eds.), *Proceedings of the 4th International Conference on Microplastic Pollution in the Mediterranean Sea*, Springer Water,
https://doi.org/10.1007/978-3-032-23847-4_36

Materials and Methods

Air sampling campaigns were conducted in five industrial facilities processing virgin polymers, recycled plastics, and undifferentiated plastic waste. The sampling design encompassed multiple workplace microenvironments, including high-stress production areas (e.g., cutting, grinding, shredding), secondary processing zones, office spaces, personal samplers worn by operators during their shifts, and outdoor locations to assess ambient background.

Airborne particles were collected on two types of filters: polycarbonate filters for particle counting and morphometric analysis, and quartz filters for polymer mass quantification.

Polycarbonate filters were subjected to oxidative digestion using Fenton's reagent, followed by Nile Red staining and acetone vapor clarification, enabling automated fluorescence microscopy for the detection, visualization, and morphometric profiling of particles in the size range from a few micrometers up to several hundred micrometers.

Quartz filters were analyzed by Py-GC-MS, which allowed for polymer-specific identification and quantitative assessment of airborne plastic mass (Pomata et al. 2024). Rigorous quality assurance and quality control procedures were implemented, including the use of procedural blanks, reagent pre-filtration, and recovery tests on spiked standards to evaluate accuracy and reproducibility.

Results and Discussion

The integrated methodology demonstrated consistent detection of airborne microplastics in all investigated facilities. Fluorescence microscopy enabled reliable visualization and enumeration of particles, while Py-GC-MS provided chemical identification and mass-based quantification. The complementarity of these techniques was evident: microscopy supplied detailed information on particle abundance and morphology, whereas Py-GC-MS offered robust insights into polymeric composition.

Preliminary results from the first two facilities examined showed distinct profiles associated with the type of processing activity. In facilities handling transparent plastics, elevated levels of certain polymers were recorded in primary production zones, while facilities engaged in fixture manufacturing displayed the predominance of other polymer types across both production and ancillary areas. The differentiation between personal samplers and stationary samplers further highlighted variations between individual exposure and general workplace background.

The reliability of the integrated method was confirmed by recovery rates obtained in validation experiments, which consistently demonstrated the robustness of the workflow. These findings emphasize the importance of adopting dual analytical approaches to overcome the limitations of single techniques and provide a holistic characterization of airborne microplastics in occupational environments.

From a broader perspective, the results point to significant potential exposure of workers in plastic-processing industries, underscoring the urgency of implementing targeted control measures. In addition, the study contributes to the establishment of evidence-based safety guidelines and lays the groundwork for toxicological studies that can further elucidate the health implications of chronic inhalation of airborne microplastics.

Conclusions

This study presents the development and validation of an integrated methodology for the assessment of airborne microplastics in occupational settings. By combining fluorescence microscopy and Py-GC-MS, the approach enables simultaneous evaluation of particle abundance, morphology, and polymer composition, thus overcoming the limitations of individual techniques.

The findings confirm that plastic-processing facilities represent relevant sources of airborne microplastics and highlight the potential risks for workers. The validated protocol, supported by rigorous quality assurance, provides a robust analytical framework for future investigations. Beyond its methodological contribution, this work stresses the need for systematic monitoring and preventive strategies, ultimately supporting the development of regulatory guidelines to protect both occupational and public health.

References

Boccia P, Mondellini S, Mauro S, Zanellato M, Parolini M, Sturchio E (2024) Potential effects of environmental and occupational exposure to microplastics: an overview of air contamination. Toxics 12(5):320

Gasperi J, Wright SL, Dris R, Collard F, Mandin C, Guerrouache M, Langlois V, Kelly FJ, Tassin B (2018) Microplastics in air: are we breathing it in? Micro Nanoplastics Ed Dr Teresa AP Rocha-St 1:1–5

Jiang W, Yan X, Lv Y (2025) A critical review on the migration, transformation, sampling, analysis and environmental effects of microplastics in the environment. J Environ Sci 154:645–664

Kalaronis D, Ainali NM, Evgenidou E, Kyzas GZ, Yang X, Bikiaris DN, Lambropoulou DA (2022) Microscopic techniques as means for the determination of microplastics and nanoplastics in the aquatic environment: a concise review. Green Anal Chem 3:100036

Kochanek A, Grąz K, Potok H, Gronba-Chyła A, Kwaśny J, Wiewiórska I, Ciuła J, Basta E, Łapiński J (2025) Micro and nanoplastics in the environment: current state of research, sources of origin, health risks, and regulations—a comprehensive review. Toxics 13(7):564

Loganathan Y, Kizhakedathil MPJ (2023) A review on microplastics—an indelible ubiquitous pollutant. Biointerface Res Appl Chem 13(2):126

Pomata D, La Nasa J, Biale G, Barlucchi L, Ceccarini A, Di Filippo P, Riccardi C, Buiarelli F, Modugno F, Simonetti G (2024) Plastic breath: Quantification of microplastics and polymer additives in airborne particles. Sci Total Environ 932:173031

Zarus GM, Muianga C, Brenner S, Stallings K, Casillas G, Pohl HR, Mumtaz MM, Gehle K (2023) Worker studies suggest unique liver carcinogenicity potential of polyvinyl chloride microplastics. Am J Ind Med 66:1033–1047

Chapter 37
X-Ray Computed Tomography: A Novel Approach for Investigating Microplastics in the Fish Digestive System

Viktória Parobková, Lukáš Maleček, Gabriela Kalčíková, Michaela Vykypělová, Marcela Buchtová, Ondřej Adamovský, Tomáš Zikmund, and Jozef Kaiser

Abstract Plastic pollution is a major contributor to the global environmental crisis and remains one of the most pressing threats to ecosystems worldwide. Each year, millions of tons of plastic waste enter oceans and terrestrial landscapes, where it persists, contaminates environments, and disrupts biodiversity (Sehnal et al. Front Microbiol 12, 2021). As larger debris degrades into microplastics (MPs; 1–1000 µm) (GESAMP (2016) Sources, fate and effects of microplastics in the marine environment: part two of a global assessment. IMO, London, Rep. Stud. GESAMP No. 93, pp 220), the risks escalate, threatening ecosystems and the health of diverse organisms, including humans (Bhuyan et al. (2022) Front Environ Sci 10, 827289; Prata et al., Sci Total Environ 702, 2020).

V. Parobková (✉) · L. Maleček · T. Zikmund (✉) · J. Kaiser
Central European Institute of Technology, Brno University of Technology, Brno, Czech Republic
e-mail: viktoria.parobkova@ceitec.vutbr.cz

T. Zikmund
e-mail: tomas.zikmund@ceitec.vutbr.cz

G. Kalčíková · T. Zikmund · J. Kaiser
Faculty of Mechanical Engineering, Brno University of Technology, Brno, Czech Republic

G. Kalčíková
Faculty of Chemistry and Chemical Technology, University of Ljubljana, Ljubljana, Slovenia

M. Buchtová
Department of Experimental Biology, Faculty of Science, Masaryk University, Brno, Czech Republic

Laboratory of Molecular Morphogenesis, Institute of Animal Physiology and Genetics, Czech Academy of Sciences, Brno, Czech Republic

M. Vykypělová · O. Adamovský
RECETOX, Faculty of Science, Masaryk University, Brno, Czech Republic

G. Gentile et al. (eds.), *Proceedings of the 4th International Conference on Microplastic Pollution in the Mediterranean Sea*, Springer Water,
https://doi.org/10.1007/978-3-032-23847-4_37

Introduction

Plastic pollution is a major contributor to the global environmental crisis and remains one of the most pressing threats to ecosystems worldwide. Each year, millions of tons of plastic waste enter oceans and terrestrial landscapes, where it persists, contaminates environments, and disrupts biodiversity (Sehnal et al. 2021). As larger debris degrades into microplastics (MPs; 1–1000 μm) (GESAMP 2016), the risks escalate, threatening ecosystems and the health of diverse organisms, including humans (Bhuyan et al. 2022; Prata et al. 2020). MPs can enter organisms through inhalation or ingestion and have been detected in aquatic species as well as human tissues (Toussaint et al. 2019). Once internalized, MPs act as exogenous materials that may trigger toxicological responses, including physical obstruction, chemical leaching, and microbiological imbalances, all of which can damage surrounding tissues (Bhuyan et al. 2022). Understanding the spatial distribution of MPs within organisms is therefore essential to assess their health effects, organ-specific accumulation, and long-term risks (Akhatova et al. 2022; Oleksiuk et al. 2022). Conventional detection methods, including pyrolysis and spectroscopy, are widely used but face limitations (Fischer et al., 2017). These techniques often require extensive tissue digestion, resulting in the loss of spatial information and increasing contamination risks. In contrast, X-ray micro-computed tomography (microCT) offers a nondestructive method to visualize MPs in biological tissues in three dimensions. Unlike traditional approaches, microCT preserves spatial context and minimizes contamination. Although its use in MP detection is not yet widespread, recent studies have shown promise. MicroCT has been applied to MPs in sediments (Trusler et al. 2023) and, more recently, in fish samples (Strafella et al. 2024). However, these studies often targeted particles larger than 1 mm, which are not environmentally representative, highlighting the need to further develop microCT for ecotoxicological research. In summary, this study addresses this gap by exploring the potential of microCT as a powerful tool for detecting MPs within biological systems.

Materials and Methods

This study followed the workflow reported in our previously published article (Parobkova et al. 2025), with a concise summary provided below. Polyethylene (PE) MPs (30–110 μm; Cospheric LLC, USA) were used due to their spherical morphology, which allows their backtracking in the obtained CT data. Adult zebrafish (*Danio Rerio*) served as model organisms. Two fish received 2 mg of PE MPs suspended in 100 μL ethanol via intragastric gavage using ultrafine Pasteur pipette tips. Although the administered concentration exceeded environmentally relevant levels, this was necessary to establish the feasibility of the microCT workflow. Fish were euthanized with Tricaine Methanesulfonate (MS222, Merck), and from one sample, the intestines were dissected. Intact fish and dissected intestines were fixed in 4% formaldehyde and stored at 4 °C. All animal procedures were approved by the Czech Ministry of Education, Youth and Sports (approval no. MSMT-12088/2024-4). For sample preparation, tissues were dehydrated through a graded ethanol series (30–90%) and stained with 1% iodine in methanol to enhance soft-tissue contrast. For microCT measurements, two setups were employed: intact zebrafish were mounted in 1.5 ml PCR tubes (*Zebrafish Setup*), while dissected intestines were placed in Kapton tubes (*Guts Setup*); in both cases, samples were stabilized in agarose gel. Imaging was performed using a Nano3DX system (Rigaku, Japan) with a voxel size of 4.2 μm for the *Zebrafish Setup* and 2.1 μm for the *Guts Setup*. Segmentation and quantification of MPs were carried out in VG Studio Max 2025.2 using a neural-network-based classification workflow (Parobkova et al. 2025), enabling 3D visualization, particle sizing, and spatial distribution analysis along the gastrointestinal tract.

Results and Discussion

PE MPs administered to zebrafish, as well as dissected guts containing these particles, were successfully visualized using two experimental microCT setups (Fig. 37.1). Tissue staining enabled the visualization of internal anatomy, allowing noninvasive assessment of MP distribution within organs. PE MPs were consistently detected in both setups, appearing as distinct dark spherical structures (Fig. 37.1A', B'). This contrast is attributed to their characteristic morphology and lower density compared to surrounding tissues, which results in reduced X-ray attenuation. These results confirm the feasibility of localizing and characterizing MPs within intact or dissected biological samples by microCT. Quantitative analysis further demonstrated the reliability of this approach. The size distribution of particles detected in microCT datasets closely matched reference measurements obtained by laser diffraction (Parobkova et al. 2025), validating the accuracy of the method. The complete size distribution of administered PE MPs was determined, with the lower detection limit established at ~ 30 μm. The detection limit reported here was established using the presented CT setups and may vary with different sample sizes, reflecting the inherent trade-off between field of view and spatial resolution. In general, smaller samples allow higher resolution and thus improved detection sensitivity.

Importantly, microCT also enabled 3D visualization of MPs within intestinal tissues (Fig. 37.1A', B'). This allowed not only the localization of individual particles but also the identification of potential accumulation sites. From the reconstructed datasets, MPs could be differentiated by proximity to the intestinal villi or lumen center, and their passage along the gastrointestinal tract could be tracked by measuring the distance from the anterior intestinal region to the particle's location (Parobkova et al. 2025). Overall, these findings establish microCT as a powerful tool for studying the 3D distribution of MPs in biological tissues and provide a

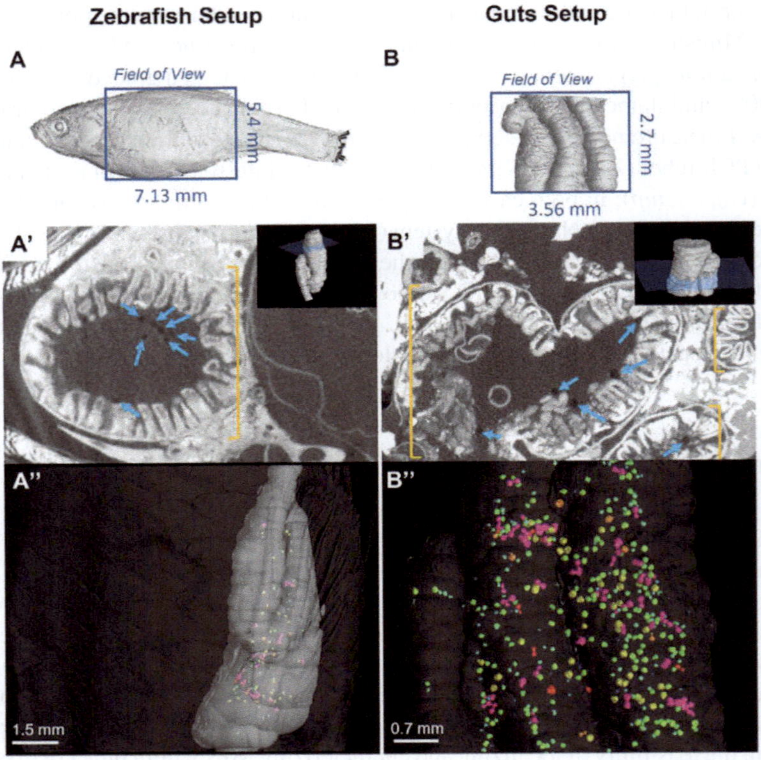

Fig. 37.1 Analysis of zebrafish and gut samples spiked with PE MPs (30–110 μm). Fields of view are shown for each setup (A, B), with detected particles visible in microCT cross-sections (A', B'). Segmented MPs were 3D rendered together with the intestines (A", B")

methodological framework for future ecotoxicological studies. Remaining challenges include testing additional polymer types and morphologies, as well as validating the approach under environmentally relevant exposure scenarios. Nevertheless, the method already demonstrates strong potential for tracking MPs after ingestion and for advancing our understanding of their fate and biological effects.

Conclusions

This study demonstrates the applicability of microCT as a nondestructive technique for detecting and localizing MPs in zebrafish. By combining optimized scanning setups with tissue staining, we were able to visualize internal anatomy and identify MPs as small as 30 μm within both intact fish and dissected guts. The accuracy of detection was confirmed by comparison with conventional particle size analysis, underscoring the robustness of the method. Importantly, the ability to reconstruct MPs in three dimensions allowed not only their quantification but also the assessment of their spatial distribution and potential accumulation sites within the gastrointestinal tract. While microCT does not provide chemical identification of polymers, its capacity to preserve spatial information and minimize contamination makes it a valuable complement to existing analytical methods. The approach introduced here lays the groundwork for future studies aimed at exploring a broader range of particle types, shapes, and environmentally relevant exposure scenarios. Taken together, these findings establish microCT as a promising methodological framework for advancing ecotoxicological research, with the potential to improve our understanding of MP fate, behavior, and biological effects in aquatic organisms and beyond.

References

Akhatova F, Ilnur I, Gölnur F et al (2022) Nanomechanical atomic force microscopy to probe cellular microplastics uptake and distribution. Int J Mol Sci 23(2):806. https://doi.org/10.3390/ijms23020806

Bhuyan MS (2022) Effects of microplastics on fish and human health. Front Environ Sci 10:827289. https://doi.org/10.3389/fenvs.2022.827289

Fischer M, Scholz-Böttcher BM (2017) Simultaneous trace identification and quantification of common types of microplastics in environmental samples by pyrolysis-gas chromatography-mass spectrometry. Environ Sci Technol 51(9):5052–5060. https://doi.org/10.1021/acs.est.6b06362

GESAMP (2016) Sources, fate and effects of microplastics in the marine environment: part two of a global assessment. IMO, London, Rep. Stud. GESAMP No. 93, pp 220

Oleksiuk K, Karolina K-K, Agata W-Ś et al (2022) Microplastic in food and water: current knowledge and awareness of consumers. Nutrients 14(22):4857. https://doi.org/10.3390/nu14224857

Parobková V, Maleček L, Zemek M et al (2025) Advancing microplastic detection in zebrafish with micro computed tomography: a novel approach to revealing microplastic distribution in organisms. J Hazard Mater 488:137442. https://doi.org/10.1016/j.jhazmat.2025.137442

Prata JC, da Costa JP, Lopes I et al (2020) Environmental exposure to microplastics: an overview of possible human health effects. Sci Total Environ 702:134455. https://doi.org/10.1016/j.scitotenv.2019.134455

Sehnal L, Brammer-Robbins E, Wormington AM et al (2021) Microbiome composition and function in aquatic vertebrates: small organisms making big impacts on aquatic animal health. Front Microbiol 12:567408

Strafella P, Giulietti N, Caputo A et al (2024) Detection of microplastics in fish using computed tomography and deep learning. Heliyon 10(21):e39875. https://doi.org/10.1016/j.heliyon.2024.e39875

Toussaint B, Raffael B, Angers-Loustau A et al (2019) Review of micro- and nanoplastic contamination in the food chain. Food Addit Contam Part A 36(5):639–673. https://doi.org/10.1080/19440049.2019.1583381

Trusler MM, Sturrock CJ, Vane CH et al (2023) X-ray computed tomography: a novel non-invasive approach for the detection of microplastics in sediments? Mar Pollut Bull 194:115350. https://doi.org/10.1016/j.marpolbul.2023.115350

Chapter 38
Preliminary Investigation of Microplastic and Microfiber Contamination in Raw Milk from Italy: Implications for Food Safety

Serena Santonicola, Michela Volgare, Mariacristina Cocca, and Giampaolo Colavita

Abstract This study aimed to assess the presence and characteristics of microplastics (MPs) in raw milk samples collected in the Molise region (Italy). To minimize contamination from equipment, milk was obtained by manual milking from three cows over a three-month period, with one sampling per month. MPs were identified, classified, and measured through microscopic analysis, followed by chemical identification of a subsample of particles using Fourier Transform Infrared microspectroscopy. Results confirmed the presence of MPs and synthetic microfibers (MFs) at a mean concentration of 3.22 ± 2.23 particles/100 mL. Additionally, a significant amount of natural MFs was detected (4.5 MFs/100 mL). The hypothesis of MP transfer from the cow (e.g., via the mammary gland) remains unconfirmed, and the occurrence of polymers, such as cellulose and polyester, found in raw milk may be traced back to various sources throughout the farm environment. These findings underscore the need for further investigations into MP sources and mitigation strategies within the dairy supply chain.

S. Santonicola · G. Colavita
Institute for Polymers, Composites and Biomaterials, National Research Council of Italy, Naples, Italy

S. Santonicola (✉) · M. Cocca
Department of Medicine and Health Sciences "V. Tiberio", University of Molise, Campobasso, Italy
e-mail: serena.santonicola@unimol.it

M. Volgare
Department of Chemical Engineering, Materials, and Industrial Production, University of Naples Federico II, Naples, Italy

G. Gentile et al. (eds.), *Proceedings of the 4th International Conference on Microplastic Pollution in the Mediterranean Sea*, Springer Water,
https://doi.org/10.1007/978-3-032-23847-4_38

Introduction

Microplastics (MPs), defined as plastic particles smaller than 5 mm, have emerged as contaminants of growing concern due to their widespread presence in marine, freshwater, and terrestrial ecosystems as well as drinking water and the food chain (Kwon et al. 2020; Cverenkárová et al. 2021). Human exposure resulting from diet has two main sources: food contaminated with MPs from the environment and/or as a result of food manufacturing and processing and contact with packaging (Kadac-Czapska et al. 2023). Although various food products have been found to be contaminated with MPs, knowledge about many foods remains lacking. While foodstuffs such as seafood, have been well investigated, milk products remain poorly studied despite their relevance in human nutrition (Cverenkárová et al. 2021; Radulescu et al. 2024).

The presence of MPs in animal-derived food reveals how environmental pollution can contribute to food contamination, with MP levels varying by investigations due to different pollution levels, farming practices, and animal feed (Lin et al. 2022; Akanda et al. 2025). In addition, potential sources of MPs in milk may include worker clothing, and milking equipment (Di Fiore et al. 2023; Rbaibi Zipak et al. 2024). Previous studies found that milk samples were contaminated with plastic particles and microfibers (MFs) increasing from farm to processed products (Da Costa Filho et al. 2021; Santonicola et al. 2025), but limitations in distinguishing contamination sources from milking equipment versus farm environment and animal exposure remain.

Therefore, this study aimed to assess the presence and characteristics of MPs and MFs in raw milk samples collected in a cow farm located in the Molise region (Italy). To exclude the potential role of milking equipment in raw milk contamination, samples were obtained through manual milking, providing a more accurate assessment of MP levels directly from farm environment.

Materials and Methods

Milk samples were obtained by manual milking from three cows over three months, with one sampling per month. Samples were placed in glass bottles washed with prefiltered water (0.45 μm pore size cellulose acetate filters, Sartorius Stedim Biotech, Gottingen, Germany) and then transported to the laboratory under cold conditions (4–7 °C). Field blank tests were conducted at all samplings, and blank tests were also carried out to detect the potential contamination in the laboratory during the analyses. Digestion was carried out applying Santonicola et al. (2025) method with some modifications, using 30% (v/v) hydrogen peroxide (Carlo Erba, Val De Reuil, France) and incubation at 50 °C for 48 h. Raw milk samples were filtered before cooling to prevent the clogging of the filters (8 μm pore size cellulose nitrate filters, Sartorius Stedim Biotech, Gottingen, Germany). MPs and MFs were identified, classified, and measured through microscopical analysis. A subsample of particles recovered on cellulose filters was transferred to MakroPor silicon filters (Thermo Fisher Scientific, Waltham, MA, USA), followed by chemical identification using Fourier Transform Infrared (FTIR) microspectroscopy (Nicolet iN10 MX, Thermo Fisher Scientific, Waltham, MA, USA). OMNIC™ Spectra™ 2.0 software (Waltham, MA, USA) was used for the analysis of the acquired spectra, and a match of more than 70% was considered for identification.

Results and Discussion

Results showed the presence of MPs and synthetic MFs in 89% of raw milk samples at a mean concentration of 3.22 ± 2.23 particles/100 mL. Additionally, a significant amount of natural/artificial MFs was detected (4.5 MFs/100 mL). The statistical analyses showed significant differences among sampling days for MPs (Kruskal-Wallis, $p = 0.047$), while no significant difference was observed between the cows selected for the sampling and milk contamination levels [ANOVA, $F(2.6) = 0.176$, $p = 0.842$]. The identification of the chemical composition of a subset (26%) of MPs and MFs indicated the prevalence (89%) of natural/artificial cellulosic polymers, and the occurrence of polyester and polypropylene particles.

Particle length was determined using ImageJ software (release 1.43, NIH, Bethesda, MD, USA), and the dimensions were obtained from the long side. Natural and synthetic MFs with a size range of 350–2000 and ≤ 350–1000 μm, respectively, dominated the total MFs, while among fragments, particles ≤350 μm were prevalent (Fig. 38.1). No significant difference was observed between the particle length for the different sampling days (Kruskal-Wallis, $p = 0.223$).

Most of the MFs were dark (52%), while 34% of the fragments were sky blue (Fig. 38.2).

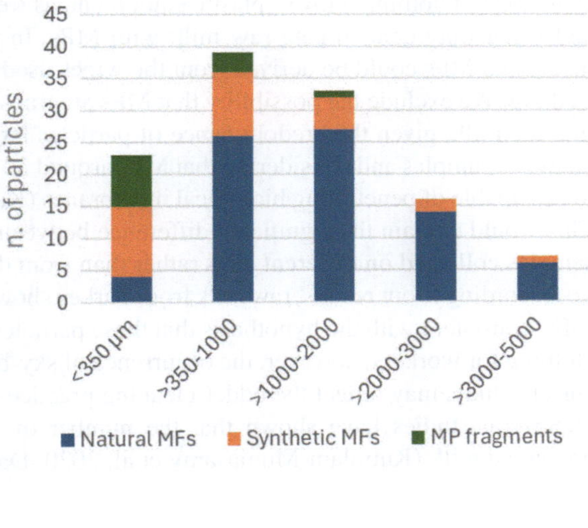

Fig. 38.1 Size distribution of the particles detected in raw milk samples

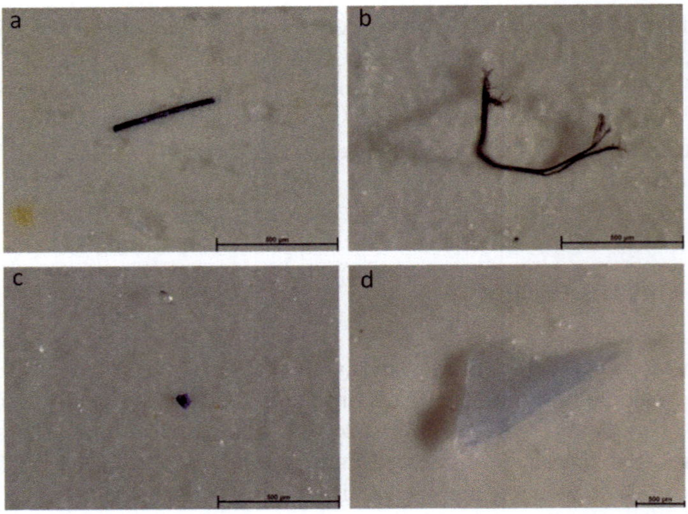

Fig. 38.2 Micrographs of coloured MFs and MPs (**a, b** synthetic and natural MFs; **c, d** fragments)

Da Costa Filho et al. (2021) evaluated the occurrence of small-sized MPs (≥ 5 μm) in raw milk collected at the milking parlor and in commercial milk. The authors found high contamination levels in all samples, ranging from 204 to 1004 MPs/100 mL. However, since the raw milk was obtained from the milking machine, it was impossible to determine whether the detected MPs originated from the cow, the milking equipment, or the milking parlor environment. Although manual milking is now uncommon, it was applied in this study to investigate contamination sources other than those associated with mechanical milking.

The occurrence of polymers such as cellulose and polyester found in raw milk may be traced back to various sources throughout the farm environment (Chakraborty et al. 2024). Clothing, gloves, plastic equipment, as well as detergents, disinfectants, and water may contaminate raw milk with MPs. In particular, the cellulose fragments and MFs could be derived from the wipes used for cleaning the udder before milking. We exclude the possibility that MPs are transferred directly from the udder into the milk, given the predominance of particles larger than 350 μm found in the analyzed samples and considering that MPs around 20 μm in size have been reported to be capable of penetrating biological membranes (Kannan and Vimalkumar 2021). This would explain the significant difference between contamination levels in milk samples collected on different days rather than from different cattle.

According to our results, raw milk from Turkey showed high levels of dark-colored MFs, consistent with the hypothesis that these particles may be shed from protective clothing for workers; however, the occurrence of sky-blue fragments, detected in the current study, may reflect the udder-cleaning practices (Rbaibi Zipak et al. 2024).

Several studies have shown that the number of MPs increases from farm to processed milk (Kutralam-Muniasamy et al. 2020; Da Costa Filho et al. 2021), and the more complex the processing and packaging process of milk, the more particles it contains (Santonicola et al. 2025). These findings provide valuable insights into the accumulation of plastic particles at different stages of the milk chain and highlight the need to properly identify contamination sources and implement appropriate control strategies and mitigation measures.

Conclusions

This study demonstrates the presence of MPs and MFs in raw milk samples obtained through manual milking. The particle size distribution (most particles ≤350 μm) excludes udder-to-milk transfer, confirming external contamination pathways. The predominance of MF particles and the significant temporal variation in contamination levels between sampling days, rather than between individual animals, indicate that contamination primarily originates from farm environment sources such as worker clothing and cleaning materials. These findings highlight the need for targeted control strategies at the farm level to minimize MPs and MFs contamination in dairy products.

References

Akanda MR, Kawsar MA, Sabuj MSS et al (2025) Microplastic pollution in livestock farming in Bangladesh: a critical review of its key findings and mitigation strategies to address public health. Environ Monit Assess 197:732

Banica AL, Radulescu C, Dulama ID et al (2024) Microplastics, polycyclic aromatic hydrocarbons, and heavy metals in milk: analyses and induced health risk assessment. Foods 13:3069

Chakraborty TK, Hasan MJ, Netema BN et al (2024) Microplastics in the commercially available branded milk in Bangladesh: an emerging threat for human health. J Hazard Mater 477:135374

Cverenkárová K, Valachovičová M, Mackuľak T et al (2021) Microplastics in the food chain. Life 11:1349

Da Costa Filho PA, Andrey D, Eriksen B et al (2021) Detection and characterization of small-sized microplastics (≥ 5 μm) in milk products. Sci Rep 11:24046

Di Fiore C, Carriera F, Russo MV et al (2023) Are microplastics a macro issue? A review on the sources of contamination, analytical challenges and impact on human health of microplastics in food. Foods 12:3915

Kadac-Czapska K, Trzebiatowska PJ, Knez E et al (2023) Microplastics in food-a critical approach to definition, sample preparation, and characterisation. Food Chem 418:135985

Kannan K, Vimalkumar K (2021) A review of human exposure to microplastics and insights into microplastics as obesogens. Front Endocrinol 12:724989

Kutralam-Muniasamy G, Pérez-Guevara F, Elizalde-Martínez I et al (2020) Branded milks–are they immune from microplastics contamination? Sci Total Environ 714:136823

Kwon JH, Kim JW, Pham TD et al (2020) Microplastics in food: a review on analytical methods and challenges. Int J Environ Res Public Health 17:6710

Lin Q, Zhao S, Pang L et al (2022) Potential risk of microplastics in processed foods: preliminary risk assessment concerning polymer types, abundance, and human exposure of microplastics. Ecotoxicol Environ Saf 247:114260

Rbaibi Zipak S, Muratoglu K, Buyukunal SK (2024) Microplastics in raw milk samples from the Marmara region in Turkey. J Consum Prot Food Saf 19:175–186
Santonicola S, Volgare M, Cocca M et al (2025) Study of fibrous microplastic and natural microfiber levels in branded milk samples from Italy. Ital J Food Saf 14:13523

Chapter 39
Chemometrics and Hyperspectral Imaging for a Standardized and Rapid Monitoring of Microplastics in the Environment

Giorgia Sciutto, Ilenia Saraceno, Emilio Catelli, Jacopo La Nasa, Greta Biale, Cristina Malegori, Silvia Prati, Francesca Modugno, and Paolo Oliveri

Introduction

Microplastic (MP) pollution is currently detected in all the (Geyer et al. 2017) main environmental compartments, from terrestrial (VaraPrasad et al. 2022) to freshwater (Wang et al. 2022) and marine systems (Carbery et al. 2018), and the associated environmental and health hazards (Betts 2008) have been the focus of intense social, scientific, and political attention. However, MP detection is nowadays still one of the biggest technological challenges, representing a new analytical issue. There is an increased need for developing standardized protocols for the sampling, quantification, and characterization of MP. These include also data treatment and visualization, which are crucial to compare different studies. Several efforts have been devoted to establish harmonized and traceable procedures. Among others, hyperspectral imaging systems represent potential cost and time effective methods, enabling a direct detection of MP when applied directly on filters without heavy sample purification and manipulations, thus avoiding potential procedural bias related to particle pre-sorting steps. Hyperspectral imaging (HSI) integrates spectral and spatial information, generating chemical maps in which each pixel contains a full spectrum. This

G. Sciutto (✉) · I. Saraceno · E. Catelli · S. Prati
Department of Chemistry, University of Bologna, Ravenna, Italy
e-mail: giorgia.sciutto@unibo.it

J. La Nasa · G. Biale · F. Modugno
Department of Chemistry and Industrial Chemistry, University of Pisa, Pisa, Italy

C. Malegori · P. Oliveri
Department of Pharmacy, University of Genova, Genoa, Italy

G. Gentile et al. (eds.), *Proceedings of the 4th International Conference on Microplastic Pollution in the Mediterranean Sea*, Springer Water,
https://doi.org/10.1007/978-3-032-23847-4_39

approach enables precise identification of chemical constituents, mapping of their spatial distribution, and detection of subtle compositional variations. The resulting multidimensional datasets, may provide robust tools for material characterization. Recently, near infrared hyperspectral imaging (NIR-HSI) has been evaluated as a technique for identifying MP directly on the filters (Piarulli et al. 2022; Vidal and Pasquini 2021). However, the use of HSI in monitoring campaigns may generate large volumes of data difficult to process and interpret. In this context, chemometric methods are emerging as essential tools for the processing and interpretation of chemical data obtained through spectroscopic techniques. These methods enable data dimensionality reduction and visualization, reducing time required for data processing and interpretation and producing graphical outcomes that can be compared among monitoring campaigns or for the integration of different results. Explorative and unmixing methods can be applied to enhance spectral features of MPs in complex matrices, identified diagnostic markers for their discrimination. To this aim, different explorative and unmixing methods have been tested and developed to determine the chemical composition to enhance the efficiency of polymer recognition. In particular, masking procedures for background removal and advanced algorithms based on multivariate data compression were explored, to enhance the spectral signal features of MP with respect to the filters used. This contribution reports the application of a HSI analysis exploiting Mid Infrared (4000–675 cm^{-1}) applied on cellulose filter, in combination with different tailored chemometric methods for data processing. In particular, an automated normalised difference image (NDI) strategy and principal component analysis (PCA) were applied and compared.

Materials and Methods

Commercially available polypropylene (PP) and polystyrene (PS) fragments, cryogenically ground from pristine pellets purchased from Merkel, with a median grain size >125 μm were used. Suspensions were prepared in a water–ethanol mixture containing a defined amount of microplastics (MPs). Each filter was loaded with approximately 0.5 mg of each polymer. The suspension was filtered by using cellulose filters (Whatman® grade 42; d = 5.5 cm; pore size = 2.5 μm) with a vacuum filter-system. As a representative case study, this work reports on the application of a micro hyperspectral imaging system (micro-HSI) operating in the mid-infrared region (4000–675 cm^{-1}) was employed for the detection of the MPs. Micro-HSI was performed in the MIR range (4000–675 cm^{-1}) by using iN10 MX infrared microscope (Thermo Scientific™ Waltham, MA, USA), equipped with an MCT-A linear array detector with a spatial resolution of 10 microns. Measurements were performed in total reflection mode, with a pixel size of 25 μm × 25 μm, step size of 25 μm × 25 μm and a spectral resolution at 16 cm^{-1}. The collected datasets were organized in a three-dimensional data matrix: the first two dimensions are vertical and horizontal spatial coordinates (pixels), while the third dimension represents the wavenumber (spectral dimension). To correct for baseline shifts and instrumental noise, spectra were converted into pseudo-absorbance, and subsequently cropped to the 3200–800 cm^{-1} range. A Savitzky–Golay first derivative (15-point window, third-degree polynomial) was then applied, followed by Standard Normal Variate (SNV) correction. Principal component analysis was performed to initially explore the dataset and to identify variables for the characterization of polymers. Mean centering was performed as the column pre-processing on the data matrice, prior to PCA. Subsequently, chemical maps were obtained by calculating the normalised difference image (NDI) (Malegori et al. 2020) according to the following formula:

$$NDI = \frac{R_{\lambda f} - R_{\lambda s}}{R_{\lambda f}}$$

where $R_{\lambda f}$ and $R_{\lambda p}$ are the reflectance values at wavelengths characteristic for the filter and the polymer. For cellulose filter the diagnostic band at 1643 cm^{-1} was considered, while bands at 1457 cm^{-1} and 1590 cm^{-1} were selected for the identification of PP and PS, respectively. For each pixel, the NDI value was encoded using a chromatic scale from blue (minimum) to red (maximum), generating an RGB chemical map that highlighted the presence of MP. Pixels associated with MP particles, characterized by the highest NDI ratios, appeared in red. Polibrush software was employed for data elaboration (Rocha de Oliveira et al. 2023).

Results and Discussion

Principal Component Analysis (PCA) identifies directions of maximum variance within datasets, assuming that higher variance corresponds to greater informational content. Principal components (PCs) are linear combinations of the original variables, with the associated coefficients, or loadings, indicating the contribution of each variable (ranging from − 1 to + 1) to the respective PC. Loadings can be visualized in two-dimensional plots to assess inter-variable correlations, while score values, representing the projections of data points onto the PC-defined space, can be displayed in scatter plots to reveal multidimensional structures such as clusters or trends. Additionally, score maps employ a chromatic scale to spatially represent score variations for each PC, providing spatial location of components. The results indicate that the PC1 (26.8% of variance explained) score map effectively discriminates MPs, characterized by green/blue hues corresponding to low score values, from cellulose, which is associated with red coloration indicative of high score values (Fig. 39.1). The variability in score values and the corresponding color range from green to dark blue suggest the presence of different types of MPs. In contrast, the PC4 (4.8% of variance explained) score map revealed red spots within a predominantly red background, potentially related to MPs but distinct from those identified in the PC1 map. After the preliminary map analysis, loading profiles were examined in order to identify the spectral bands mainly involved in the definition of each polymer. Negative values of loadings along PC1 reflected the contribution of specific variables that plays crucial roles in the characterization of PS MPs (blue spots), such as: 1450, 1590, 2923, 3040 cm^{-1}.

The evaluation of loadings on PC4 reveals the contribution of diagnostic variables ascribable to PP, that present positive loadings values and can be thus associated to the red pixels in the maps: 1160, 1373, 1457, 2962 cm^{-1}.

The most relevant variables identified with PCA are used for the NDI elaboration, where the difference between the MP and the support filter is represented by a color-scale image whose range is between the maximum (yellow) and the minimum (blue) reflectance values (Fig. 39.2). Moreover, spectra can be interactively extracted by selecting the pixels of interest from the chemical maps.

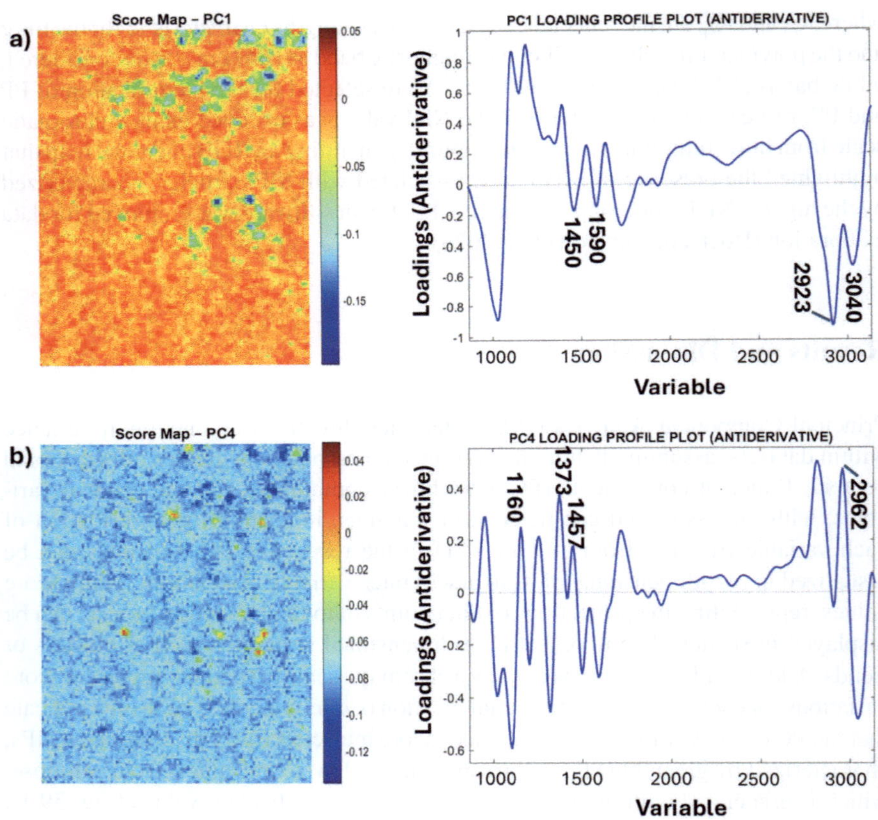

Fig. 39.1 HSI analysis of a cellulose filter containing a MPs mixture. **a** PC1 score map with the corresponding loading profile (spectra corrected using a first derivative transformation). **b** PC4 score map with the corresponding loading profile (spectra corrected using a first derivative transformation)

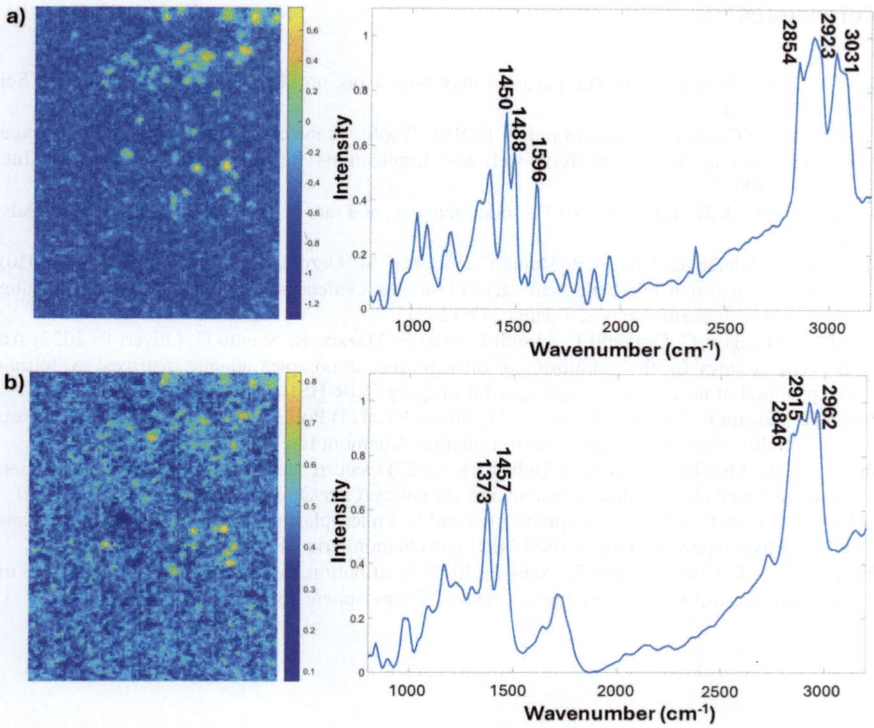

Fig. 39.2 HSI analysis of a cellulose filter containing an MPs mixture. **a** NDI image of PS with the representative spectrum extracted from selected particles; **b** NDI image of PP with the representative spectrum extracted from selected particles

Conclusions

The study demonstrates the applicability of the HSI method for the qualitative detection of MP in matrices. Beyond the acquisition phase, the establishment of harmonized and reproducible protocols necessarily requires careful attention to multivariate data processing and visualization, since the comparability of results across studies ultimately depends on these steps. In this sense, the integration of acquisition, analysis, and visualization workflows not only supports the definition of standardized procedures but also paves the way for robust strategies in comparative intra-laboratory studies and for the development of reliable, transferable methodologies.

References

Betts K (2008) Why small plastic particles may pose a big problem in the oceans. Environ Sci Technol 42(24)

Carbery M, O'Connor W, Palanisami T (2018) Trophic transfer of microplastics and mixed contaminants in the marine food web and implications for human health. Environ. Int. 115:400–409

Geyer R, Jambeck JR, Law KL (2017) Production, use, and fate of all plastics ever made. Sci Adv 3(7)

Malegori C, Alladio E, Oliveri P, Manis C, Vincenti M, Garofano P, Barni F, Berti A (2020) Identification of invisible biological traces in forensic evidences by hyperspectral NIR imaging combined with chemometrics. Talanta 215:120911

Piarulli S, Malegori C, Grasselli F, Airoldi L, Prati S, Mazzeo R, Sciutto G, Oliveri P (2022) An effective strategy for the monitoring of microplastics in complex aquatic matrices: exploiting the potential of near infrared hyperspectral imaging (NIR-HSI). Chemosphere 286:131861

Rocha de Oliveira R, Malegori C, Sciutto G, Oliveri P (2023) PoliBrush—a user-friendly software to aid multivariate image analysis dissemination. Chemom Intell Lab Syst 240:104918

VaraPrasad K, Mondal A, Cheela V, Dubey BK (2022) Occurrences, impacts and characterization of microplastics in terrestrial ecosystem to aid policy. Curr Opin Environ Sci. Health 100361

Vidal C, Pasquini C (2021) A comprehensive and fast microplastics identification based on near-infrared hyperspectral imaging (HSI-NIR) and chemometrics. Environ Pollut 285:117251

Wang Y, Zhou B, Chen H, Yuan R, Wang F (2022) Distribution, biological effects and biofilms of microplastics in freshwater systems-A review. Chemosphere 134370

Chapter 40
Green Biomass-Derived Porous Systems: Multifunctional and Lightweight Materials

Alfredo Sorrentino, Rachele Castaldo, Roberto Avolio, Federico Carosio, Antonia Cerbone, Mariacristina Cocca, Maria Emanuela Errico, Noemi Faggio, Lorenza Maddalena, Federico Olivieri, and Gennaro Gentile

Abstract This study aims to explore strategies for the valorization of lignocellulosic biomasses to develop multifunctional hierarchically porous materials by incorporating micro/mesoporous particles into macroporous foams. Macroporous aerogels were fabricated using ball-milled cellulose and rice husk and then by inducing their crosslinking using citric acid, yielding lightweight materials with good water stability. In parallel, micro/mesoporous cellulose-based polymers were developed via functionalization strategies to enhance adsorption toward a broad range of organic pollutants. The integration of these two systems resulted in hierarchically porous materials with accessible multi-scale porosity. The resulting materials exhibit low density, tunable shapes, water stability, improved thermal stability, high adsorption capacity for organic contaminants.

Introduction

Following the rapid growth of the population and the development of society, there has been an increasing need for resources to sustain such growth. The growing use of resources leads to two major issues: resource depletion and increased waste production, with consequent environmental impacts. Suffice it to say that, in 2016 alone,

---

A. Sorrentino (✉) · R. Castaldo · R. Avolio · A. Cerbone · M. Cocca · M. E. Errico · N. Faggio · F. Olivieri · G. Gentile
Institute of Polymers, Composites and Biomaterials, National Research Council of Italy, Naples, Italy
e-mail: alfredosorrentino@cnr.it

F. Carosio · L. Maddalena
Department of Applied Science and Technology, Politecnico di Torino, Alessandria, Italy

A. Cerbone
Department of Chemical Sciences, University of Naples Federico II, Naples, Italy

G. Gentile et al. (eds.), *Proceedings of the 4th International Conference on Microplastic Pollution in the Mediterranean Sea*, Springer Water,
https://doi.org/10.1007/978-3-032-23847-4_40

2.01 billion tonnes of solid waste were produced (Sharma et al. 2020), resulting in significant challenges related to waste management (Pfaltzgraff et al. 2013). One of the strategies to address both resource depletion and waste management is the recovery and valorization of waste, particularly biomass waste (especially lignocellulosic biomass). A large portion of these materials can be obtained from by-products of the agri-food industry, including rice husk (RH), sugarcane leaves (SCL), sugarcane bagasse (SCB), corn husk (CH), barley husk (BH), coconut shell (CS), peanut shell (PS), coffee silverskin (CS) and wheat straw (WS) residues, among others (Sharma et al. 2020; Capuano et al. 2024).

These materials can be processed to obtain products that can be reused for other applications, including the production of cellulose, through various techniques such as ball milling, which can yield a network of micro- and nanofibrillated cellulose fibers (Avolio et al. 2012).

In addition, specific functionalities can be imparted to cellulosic materials through appropriate strategies.

For example Marcioni et al. developed a sustainable strategy to mitigate the fire hazard of cellulose-filled rigid polyurethane foams by pre-functionalizing ball-milled cellulose through a layer-by-layer (LbL) assembly of chitosan and sodium hexametaphosphate (Marcioni et al. 2024). The optimized three-bilayer coating significantly enhanced cellulose char formation and imparted self-extinguishing to non-igniting behavior at the filler level. When incorporated into polyurethane foams (PUF) at 15 phr, the coated cellulose reduced cell size and thermal conductivity while markedly improving fire performance, achieving a 22% reduction in peak heat release rate and a 32% decrease in total smoke release compared to neat foams. The improved behavior was attributed to the formation of a dense and continuous char layer during combustion, demonstrating the effectiveness of LbL-functionalized biofillers in enabling flame-retardant polyurethane foam biocomposites without compromising sustainability.

With another aim Maglione et al. developed a highly bio-based hyper-crosslinked adsorbent by grafting poly(vinylbenzyl chloride) (PVBC) onto cellulose nanofibrils (CNF) followed by Friedel–Crafts alkylation, generating a micro/mesoporous shell with a specific surface area of ~409 $m^2\ g^{-1}$ and ~ 50% microporosity. The resulting material exhibited high adsorption performance toward 2,4-dichlorophenol (DCP), achieving up to 90% removal from dilute solutions and a maximum adsorption capacity of 284 mg g^{-1} at 298 K. Adsorption occurred via spontaneous, exothermic multilayer physisorption driven by π–π interactions, hydrogen bonding, and permanent porosity, while excellent reusability (>98% efficiency over five cycles), pH stability, and low interference from natural organic matter were observed (Maglione et al. 2025).

Results and Discussion

In this study, two porous systems with different porosity scales were developed for subsequent integration into a hierarchically porous material: a macroporous aerogel and a micro/mesoporous polymer.

Macroporous Aerogels

Macroporous aerogels based on cellulose and rice husk were obtained through an eco-friendly approach combining planetary ball milling and natural crosslinking with citric acid. Water was used to induce defibrillation of the cellulose fibrils, and ball milling conditions—including biomass-to-sphere ratio, water content, and mixing rate—were optimized to enhance fibrillation efficiency. Cellulose fibers (CF) and rice husk (RH) were treated under these optimized planetary ball milling conditions, producing well-dispersed fibrils suitable for aerogel formation. The resulting aerogels exhibited low density, tunable shapes, and water stability, making them promising candidates for sustainable porous materials.

Micro/Mesoporous Polymers

Cellulose inherently shows adsorption capabilities toward cationic species due to interactions between positively charged adsorbates and negatively charged hydroxyl groups. To broaden the adsorption range to include positively charged, negatively charged, and neutral aromatic organic pollutants, functionalization strategies were applied to develop high-surface-area micro/mesoporous cellulose-based materials. These modifications introduced additional adsorption mechanisms beyond hydrogen bonding, facilitating multilayer adsorption within the micro/mesopores and increasing the accessible adsorption volume. Lignocellulosic materials such as rice husk and hemp were similarly processed to obtain powders that could be easily incorporated into the macroporous aerogels.

Hierarchically Porous Materials

Incorporating the micro/mesoporous particles into the macroporous foams yielded hierarchically porous materials that are easy to handle, water-stable, and possess accessible micro/mesoporosity for adsorption of gases and organic pollutants from solutions. These materials exhibit low density and thermal conductivity comparable to commercial insulating panels, along with considerable adsorption capacity toward volatile organic compounds (VOCs) and other organic pollutants.

Moreover, the aromatic and extensively crosslinked structure of the micro/mesoporous filler confers thermal stability, preventing melting or ignition. This characteristic is expected to impart fire safety to the overall hierarchically porous material, a feature that will be further investigated in future studies.

Conclusions

This study demonstrates that combining macroporous aerogels with micro/mesoporous cellulose-based polymers enables the fabrication of hierarchically porous materials with low density, tunable shapes, water stability, and high adsorption capacity toward organic pollutants. The materials also exhibit thermal stability, suggesting potential for fire-safe applications. These findings highlight the potential of sustainable lignocellulosic biomass as a versatile precursor for multifunctional porous materials in environmental and insulation applications.

Akcnowledgements The author acknowledge financial support under the National Recovery and Resilience Plan (NRRP), Mission 4, Component 2, Investment 1.1, Call for tender No. 1409 published on 14.9.2022 by the Italian Ministry of University and Research (MUR), funded by the European Union—NextGenerationEU—Project Title Green cellulose bAsed fire safe and lightweight Insulating mAterials (GAIA)—CUP project (CUP: E53D23017860001- Grant Assignment Decree No. 1389 adopted on 01/09/2023 by the Italian Ministry of Ministry of University and Research (MUR).

The authors also acknowledge financial support under the Italian Science Fund (FIS 2 Call) funded by the Italian Ministry of University and Research (MUR) for the project "HIerarchically Porous materials with tailored ADsorption properties to face environmental pollution" (HIPAD, Proposal code FIS-2023-01686, CUP B53C24009520001, Grant Assignment Decree No. n. 23,314 adopted on 11/12/2024 by the Italian MUR).

References

Avolio R, Bonadies I, Capitani D, Errico ME, Gentile G, Avella M (2012) A multitechnique approach to assess the effect of ball milling on cellulose. Carbohydrate Polym 87(1):265–273

Capuano R, Avolio R, Castaldo R, Cocca M, Gentile G, Cirillo T, Nolasco A, Errico ME (2024) Up-cycling coffee silverskin into biobased functional coatings. J Clean Prod. https://doi.org/10.1016/j.jclepro.2024.143063

Maglione A, Olivieri F, Avolio R, Castaldo R, Cocca M, Errico ME, Ambrogi V, Gentile G (2025) Hyper-cross-linked cellulose nanofibrils with spontaneous and reversible adsorption of aromatic pollutants from water as a valid alternative to fossil-based adsorbents. ACS Appl Mater Interfaces. https://doi.org/10.1021/acsami.5c05009

Marcioni M, Maddalena L, Avolio R, Castaldo R, Gentile G, Carosio F (2024) Layer-by-layer coated cellulose reduces the fire risk of polyurethane foam biocomposites. Polym Degrad Stability. https://doi.org/10.1016/j.polymdegradstab.2024.110935

Pfaltzgraff LA, Cooper EC, Budarin V, Clark JH (2013) Food waste biomass: a resource for high-value chemicals. https://doi.org/10.1039/C2GC36978H
Sharma G, Kaur M, Punj S, Singh K (2020) Biomass as a sustainable resource for value-added modern materials: a review. Green Chem. https://doi.org/10.1002/bbb.2079

Chapter 41
Unmasking the Resistome on Microplastics in Marine Habitats: Potential Routes of Antibiotic Resistance to Humans via Shellfish

Eloísa Toscan Massignam, Giulia Baggio, Arianna Peruzzo, Andrea Di Cesare, Raffaella Sabatino, Vanessa Moschino, Fabiana Corami, Maria Gabriella Marin, Silvia Galafassi, and Carmen Losasso

Eloísa Toscan Massignam and Giulia Baggio: Co-first authorship

E. T. Massignam · G. Baggio · A. Peruzzo · C. Losasso (✉)
Istituto Zooprofilattico Sperimentale Delle Venezie, Legnaro, PD, Italy
e-mail: closasso@izsvenezie.it

E. T. Massignam · A. Peruzzo
National Programme in One Health Approaches to Infectious Diseases and Life Science Research, Department of Public Health, Experimental and Forensic Medicine, University of Pavia, Pavia, Italy

A. Di Cesare · R. Sabatino · S. Galafassi
Water Research Institute, National Research Council, Verbania, Italy

V. Moschino
Institute of Marine Sciences, National Research Council, Venice, Italy

F. Corami
Institute of Polar Sciences, National Research Council, Campus Scientifico-Ca' Foscari University of Venice, Venezia-Mestre, Italy

Department of Environmental Sciences, Informatics, and Statistics, C, Foscari University of Venice, Venezia-Mestre, Italy

M. G. Marin
Department of Biology, University of Padova, Padova, Italy

G. Gentile et al. (eds.), *Proceedings of the 4th International Conference on Microplastic Pollution in the Mediterranean Sea*, Springer Water,
https://doi.org/10.1007/978-3-032-23847-4_41

Introduction

Recent studies provide strong evidence that humans are widely exposed to microplastics (MPs; size range 5 mm–1 μm) dispersed throughout the environment (Noventa et al. 2021). Given their widespread presence in the marine environment, seafood—an essential component of the human diet—represents a critical route of exposure (Carbery et al. 2018). Predictive studies estimate that human ingestion of MPs through mussel consumption may range from 123 particles per person per year to as many as 4620 particles per p-erson per year in countries with high shellfish intake (Catarino et al 2018). Besides the toxic effects of plastic-derived chemicals, MPs in aquatic ecosystems provide an ideal substrate for biofilm development. These biofilms host diverse microbial communities, including human pathogens, and facilitate gene transfer processes such as the dissemination of virulence and antimicrobial resistance genes (ARGs) (Wu et al. 2019). This specific interaction between MPs and microbial biofilms gives rise to a unique ecological niche known as the plastisphere (Zettler et al. 2013). The plastisphere can persist for long periods, acting as a reservoir for various contaminants while supporting a complex and rich microbiota. Studies have reported the presence of diverse eukaryotic and bacterial organisms on MPs, including protozoa, algae, fungi, cyanobacteria, diatoms, ciliates, hydroids, and heterotrophic bacteria. These organisms interact through trophic, symbiotic, and predatory relationships, forming a small but unique ecological system distinct from microbial communities colonizing other natural particles in the same environment (Zettler et al. 2013, Zheng et al. 2023). Evidence indicates that MPs create favorable conditions for the persistence and proliferation of antimicrobial-resistant bacteria (ARB), with resistance genes being transferred to pathogenic strains via horizontal gene transfer (HGT) (Zettler et al. 2013; Zheng et al. 2023; Arias-Andres et al. 2019). ARB concentrations have been found to be 100–5000 times higher on MPs surfaces compared with the surrounding seawater, thereby identify the plastisphere as a potential hotspot for antimicrobial resistance selection and dissemination (Zhang et al. 2020). Despite these concerns, knowledge about the plastisphere and its potential risks is still quite fragmented. Many of the studies stressed the need to provide a more comprehensive overview of the resistome. Consequently, while evidence strongly suggests that MPs act not only as chemical pollutants but also as vectors of microbial risks, available data are still insufficient to perform a complete risk assessment of their impact on human health, particularly in relation to microbiological hazards (Sathicq et al. 2021, Bocci et al. 2024).

In this study, we investigated the resistome associated with microplastic particles collected from a marine coastal area in the northern Adriatic Sea, a region used for commercial bivalve mollusk harvesting. We characterized the polymer composition of the particles using Fourier-transform infrared spectroscopy (FTIR) and applied a metagenomics and a metataxonomic approach to explore ARGs and the associated microbial communities.

Materials and Methods

The MPs under study were sampled in areas of the North Adriatic Sea. Specifically, the sampling was carried out using specific sampling nets, towed by the vessel in the coastal area in front of Pellestrina Island (Venice/Italy). The mesh opening of the net (manta) was about 300 μm (300–333 μm). The net was deployed on one side of the vessel at a distance of 3–4 m, i.e., outside the wake zone, to avoid interference from the turbulent water flow. The navigation speed was 2–3 knots for a duration of 30 min.

Each sample was analyzed under an optical microscope in order to visually select the MPs present. The obtained MPs were characterized for the chemical composition by FTIR. For the molecular analysis, total DNA from MPs were extracted by commercial kit—DNA Mini Kit (QIAGEN, Hilden, Germany), following the manufacturer's instructions. In particular, thermal lysis of the bacterial component present in the matrix under examination was carried out at 56 °C for 2 h, and RNase A (100 mg/mL) was added to each sample to ensure a preparation free from RNA contamination. The total DNA was resuspended in 200 μL of nuclease-free water and stored at −20 °C.

We applied a metataxonomic and metagenomics approach to characterize the resistome and the interactions between ARGs and microbial community. For metataxonomic approach, each obtained DNA sample was used as a template for amplification of the V3-V4 hypervariable regions of the 16S rRNA gene. The 16S rRNA gene library was prepared according to the Illumina 16S Metagenomic Sequencing Library Preparation protocol, following the instructions of the Nextera XT Library Prep kit (Illumina Inc., San Diego, CA, USA). Libraries were checked for both concentration and quality using Qubit dsDNA 1X High Sensitivity and 2200 TapeStation (Agilent), respectively. Samples were equimolarity pooled, and sequencing was performed with an Illumina MiSeq platform using a MiSeq 600V3 cartridge (600 cycles, 2 × 300 bp, paired ends reads). The raw data obtained were filtered and trimmed. Then, the forward and the reverse reads were merged to obtain the full denoised sequences. SILVA ribosomal database was used to assign taxonomy to the identified Amplicon Sequence Variants (ASVs) (Quast et al. 2012). The pathobiome was investigated by filtering of ASV table to keep only the ASVs belonging to potentially pathogenic bacterial genera, as described in Bartlett et al. (2022).

For the metagenomics shotgun approach, a subset of three samples were underpinned to shotgun sequencing. The subset of samples was obtained by pooling together DNA samples with similar microbial communities. Libraries were prepared using the Nextera XT DNA Library Preparation Kit (Illumina INC., San Diego; CA, USA). The average library size was assessed with a 2200 TapeStation system (Agilent). Samples were equimolarity pooled and diluted until 1.3 pM. Sequencing was carried out on the Illumina NextSeq 550 platform (paired-end, 2 × 150 bp). The shotgun sequencing raw data were processed using Trimgalore, using a Phred score cutoff of 20 and keeping only sequences longer than 100 bp. The resistome was extracted from the trimmed reads using Resistance Gene Identifier (RGI), with the CARD database as a reference (Alcock et al. 2023)

Results and Discussion

During the marine campaigns, five environmental MPs samples were collected. A total of nine MPs were selected for chemical composition characterization and molecular analysis. These MPs were identified as polyethylene (HDPE and LDPE), polypropylene, polystyrene, polyolefin, and composite materials. The measuring of the particle sizes was ranging from 1 μm to over 5 mm and classified them as fibers, fragments, and sheets.

Amplicon sequencing of identified MPs revealed a diverse microbial community, with 1198 unique ASVs detected after quality filtering. At the family level, the most abundant taxa were Flavobacteriaceae, Pseudoalteromonadaceae, Vibronaceae, Alteromonadaceae and Alcanivoracaceae. Interestingly, several families, including Listeriaceae, Staphylococcaceae, Enterococcaceae and Desulfobacteraceae, were detected on MPs but are not commonly reported in marine water microbial surveys, suggesting that MPs may select for a distinct microbial signature. Pathobiota analysis showed that potentially pathogenic taxa accountant for 5% of the total microbial community, with Vibrio as the dominant colonizing genus. Other minor potentially pathogenic genera were also detected (such as Psychrobacter and Pseudomonas), reflecting the capacity of MPs to act as a reservoir for human pathogen in the marine environment.

Shotgun metagenomics sequencing revealed a diverse resistome associated with microplastic particles collected in the northern Adriatic Sea. A total of 245 different ARGs were identified, spanning multiple classes. The resistome was dominated by genes that conferring resistance to aminoglycoside antibiotic (41.8%) and fluoroquinolones (36.7%), followed by macrolides (7.6%), peptides (2.6%), glycopeptide (2.3%), diaminopyrimidines (1.4%) and phenicols (1.2%). Other ARGs, including those conferring resistance to tetracyclines, carbapenems, nitromidazoles, penams, lincosamides, and rare classes, accounted collectively for less than 2% of the total. These finding indicate that MPs act as reservoirs for clinically relevant and environmental ARGs, with a strong enrichment of genes associated with human and veterinary antibiotic usage. The results highlight the potential role of MPs as vectors for the dissemination of antimicrobial resistance in marine environments, raising implication for human health via seafood consumption.

Conclusions

The results clearly demonstrate that microplastic particles collected in the marine environment not only exhibit diverse chemical compositions but also function as important substrates for microbial colonization. The presence of a heterogeneous microbial community, including families not commonly reported in marine water surveys, reinforces the hypothesis that MPs act as selective niches capable of sustaining a distinct microbiome. The detection of potential pathogens, such as Vibrio and other clinically relevant genera, highlights the role of MPs as reservoirs for microorganisms with direct implications for human health.

In parallel, the characterization of the resistome associated with MPs revealed a wide diversity of ARGs, dominated by those linked to human and veterinary use. This finding confirms that MPs may contribute to the dissemination of antimicrobial resistance in marine environments, representing an additional risk of transfer through the food chain via seafood consumption.

Therefore, the co-occurrence of distinct microbial communities, potential pathogens, and clinically relevant ARGs positions MPs as ecological and sanitary hotspots, underscoring the need for continuous monitoring and mitigation strategies that address not only the environmental impact of plastics but also their direct implications for public health and food safety.

References

Alcock et al (2023) CARD 2023: expanded curation, support for machine learning, and resistome prediction at the comprehensive antibiotic resistance database. Nucleic Acids Res 51:D690–D699 [PMID 36263822]

Arias-Andres M et al (2019) Collateral effects of microplastic pollution on aquatic microorganisms: an ecological perspective. Trac Trends Anal Chem 112:234e240

Bartlett A et al (2022) A comprehensive list of bacterial pathogens infecting humans. Microbiology 168(12):001269

Bocci V et al (2024) Freshwater plastisphere: a review on biodiversity, risks, and biodegradation potential with implications for the aquatic ecosystem health. Front Microbiol Sec Aquatic Microbiol 15

Carbery M et al (2018) Trophic transfer of microplastics and mixed contaminants in the marine food web and implications for human health. Environ Int 115:400–409

Catarino AI et al (2018) Low levels of microplastics (MP) in wild mussels indicate that MP ingestion by humans is minimal compared to exposure via house-hold fibres fallout during a meal. Environ Pollution 237:675e684

Noventa S et al (2021) Paradigms to assess the human health risks of nano-and microplastics. Microplastics Nanoplastics 1(1):1–27

Quast C et al (2012) The SILVA ribosomal RNA gene database project: improved data processing and web-based tools. Nucleic Acids Res 41(D1)

Sathicq MB et al (2021) Are microplastic particles a hotspot for the spread and the persistence of antibiotic resistance in aquatic systems? Environ Pollut 279:116896

Wu X et al (2019) Selective enrichment of bacterial pathogens by microplastic biofilm. Water Res 165:114979

Zettler ER et al (2013) Life in the “Plastisphere”: microbial communities on plastic marine debris. Environ Sci Technol 7137–7146

Zhang Y et al (2020) Potential risks of microplastics combined with superbugs: enrichment of antibiotic resistant bacteria on the surface of microplastics in mariculture system. Ecotoxicol Environ Saf 187:109852

Zheng Z et al (2023) Interaction between microplastic biofilm formation and antibiotics: effect of microplastic biofilm and its driving mechanisms on antibiotic resistance gene. J Hazard Mater 459:132099

Chapter 42
Origin, Fate and Transport of Microplastics in the Atlantic Region Down to 1200 m Depth

Daura Vega-Moreno, Marina Vargas-Ferraz, Álvaro Cubas-Viera, Ana Molina-Rodríguez, Lianela Peña-García, Bárbara Abaroa-Pérez, Eugenio Fraile-Nuez, Javier Hernández-Borges, and Francisco Machín

Introduction

Although in recent years there has been a slight increase in papers published on subsurface samples, very few studies so far have characterized the presence and composition of microplastics (MPs) in the water column below 50 m. These studies report the presence of small-sized MPs (SMP) throughout the water column, even at depths below 1000 m (Choy et al. 2019; Cincinelli et al. 2019; Egger et al. 2020). The main hypothesis is that these SMPs are found at such depths due to their vertical sinking from the surface, according to their density and physicochemical characteristics (Chubarenko and Stepanova 2017; Eo et al. 2021; Erni-Cassola et al. 2019;

D. Vega-Moreno (✉) · M. Vargas-Ferraz · A. Molina-Rodríguez · L. Peña-García
OpenPLAS GroupChemistry Department, University of Las Palmas de Gran Canaria (ULPGC), Las Palmas de Gran Canaria, Spain
e-mail: daura.vega@ulpgc.es

Á. Cubas-Viera · F. Machín
Physical Oceanography and Applied Geophysics (OFYGA), IU-ECOAQUA, University of Las Palmas de Gran Canaria (ULPGC), Las Palmas de Gran Canaria, Spain

B. Abaroa-Pérez
Marine Litter Observatory (OBAM), La Oliva Town Hall and Fuerteventura Island Council, Las Palmas, Spain

E. Fraile-Nuez
Oceanographic Center of the Canary Islands, Spanish Institute of Oceanography, Superior Council of Scientific Research (CSIC) (IEO-CSIC), Santa Cruz de Tenerife, Spain

J. Hernández-Borges
Department of Chemistry, Departmental Unit of Analytical Chemistry, Faculty of Sciences, University of La Laguna (ULL), San Cristóbal de La Laguna, Spain

G. Gentile et al. (eds.), *Proceedings of the 4th International Conference on Microplastic Pollution in the Mediterranean Sea*, Springer Water,
https://doi.org/10.1007/978-3-032-23847-4_42

Kukulka et al. 2012). Although some authors highlight the relevance of other parameters, such as particle shape and size, in the vertical transport (Poulain et al. 2019), some others suggest that transport mechanisms other than simple buoyancy must be involved (Morales-Caselles et al. 2021). In addition, there is also evidence that transport mechanisms other than buoyancy-driven sinking may account for the presence of SMPs at these depths. One example is the detection of low-density polymers such as polyethylene and polypropylene at depths exceeding 4000 m (Li et al. 2020; Peng et al. 2018), despite being classified as "buoyant microplastics". The second is that vertical MPs profiles in the water column are not homogeneous, with concentrations varying by depth and in some cases exceeding surface levels (Choy et al. 2019; Vega-Moreno et al. 2021, 2024).

Materials and Methods

To characterize the salinity distribution in the North Atlantic, we used the Argo-based global climatology developed by Roemmich and Gilson (available at https://sio-argo.ucsd.edu/RG_Climatology.html). This climatology integrates Argo float observations and provides monthly temperature and salinity fields at the global scale. From this dataset, we extracted the salinity values at 1000 m depth for November 2020, corresponding to the sampling period of the nanoplastic survey conducted by ten Hietbrink et al. (2025). This allowed us to generate a map of isohalines representative of the hydrographic conditions at intermediate depth during that month.

Results and Discussion

The @OpenPLAS group (led by Dr. Vega-Moreno) has been working since 2018 on the study of MPs in the Atlantic region down to 1200 m depth, focusing their research so far on the eastern North Atlantic, near the transition zone between the Mediterranean and Atlantic waters. In a study published in 2024 on oceanic samples near the Canary Islands (28° N), @OpenPLAS described the link between significantly high concentrations of MPs at 1100 m depth and the presence of Mediterranean Water (MW). This water mass can be detected by its relatively high salinity signature in the eastern North Atlantic between 800 and 2000 m, with a clear core located at some 1100 m depth (Iorga et al. 1999; Iorga and Lozier 1999). The analysis documented by @OpenPLAS in 2024 in the Canary Islands region highlights the fact that this water mass transports SMP as passive tracers of the flow (in that article, MPs of around 100 μm in size). Moreover, based on a recent study published in *Nature* (ten Hietbrink et al. 2025), we believe that this relationship with MW might also apply to nanoplastics. These authors analyze the concentration of nanoplastics in the eastern North Atlantic at the surface, near the seafloor, and at 1000 m depth. This last depth is precisely where the MW is located, so this coincidence prompted us to reinterpret such findings in light of our results on SMPs.

Figure 42.1 presents the isohalines at 1000 m depth. The influence of the MW is evident, originating at the Strait of Gibraltar and leaving a high-salinity footprint that extends mainly northward and westward. As this water mass spreads through the Atlantic, it progressively mixes with the surrounding waters, but its trajectory and extent are still recognizable. The figure also shows the stations where nanoplastic at 1000 m depth have been gathered (ten Hietbrink et al. 2025). Figure 42.2 compiles the analyzed nanoplastic concentrations (sum of all polymer types). The highest concentration appears at station 8, followed by stations 1, 2, 3, and 7, which show similar concentrations among themselves. In contrast, stations 4, 5, 6, and 9 show the lowest nanoplastic concentrations. If we combine Figs. 42.1 and 42.2, we can see that stations 1 through 8, i.e., those with higher salinities at 1000 m (35.34–35.53) and therefore a greater contribution of MW, coincide with those where the highest nanoplastic concentrations have been obtained, while those with salinities below 35.29 show lower nanoplastic levels. That finding mirrors @OpenPLAS observations for microplastic samples near the Canary Islands at 1100 m depth.

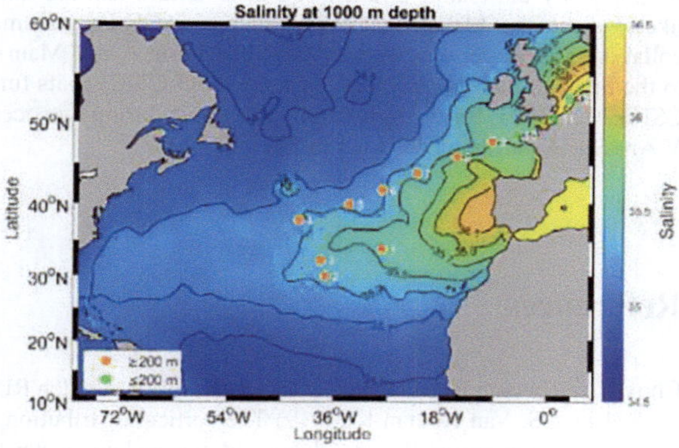

Fig. 42.1 Salinity map with stations sampled

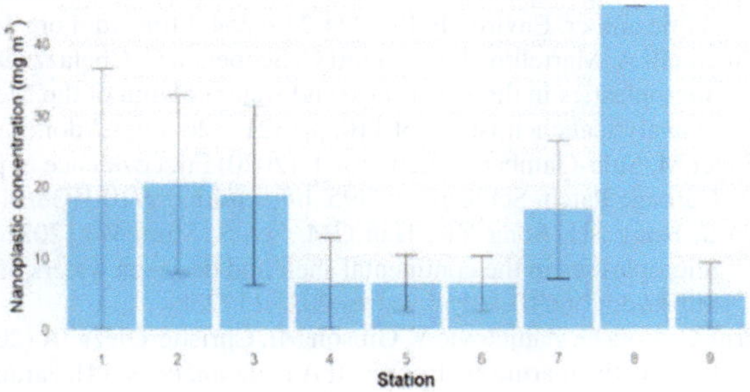

Fig. 42.2 Nanoplastic concentration at those stations at 1000 m depth in Nov 2020 (data from ten Hietbrink et al. 2025)

Conclusions

Our results confirm that microplastics are not restricted to the ocean surface but are also present at intermediate depths of the eastern North Atlantic. Elevated concentrations at ~ 1000–1100 m coincide with the core of Mediterranean Water (MW), suggesting that this water mass acts as an efficient pathway for their horizontal transport across the basin. The comparison between @OpenPLAS observations on SMPs near the Canary Islands and the nanoplastic dataset of ten Hietbrink et al. (2025) further supports this link, as the stations most influenced by MW consistently exhibit the highest plastic concentrations. These findings underline the importance of incorporating physical oceanography into microplastic research. Intermediate water masses, such as MW, can serve as effective dispersal vectors, redistributing particles far from their sources and across large oceanic regions.

Acknowledgements The authors would like to thank the Promotur Turismo Canarias Agreement for the joint implementation of initiatives aimed at intelligent management and the development of the Blue Tourism product. These efforts fall within the framework of the Recovery, Transformation, and Resilience Plan (Next Generation EU Funds), through the Tourism Sustainability Plans of the Recovery Plan. We would like to thank the collaboration of Marine Litter Observatory (OBAM), supported by the Island Council of Fuerteventura and the Oliva Town Council. We would also like to thank the Africa Canarias Challenge (ACC Programme) for supporting this work in the collaboration between the Canary Islands, Morocco, and Mauritania. We would also like to thank to the Spanish Institute of Oceanography (IEO-CSIC) in its funded project VULCANA-III (IEO-CSIC-2021-2023) for the opportunity to sample during two oceanographic cruises on board the R/V Ángeles Alvariño, carried out in 2021 and 2022.

References

Choy CA, Robison BH, Gagne TO, Erwin B, Firl E, Halden RU, Hamilton JA, Katija K, Lisin SE, Rolsky CS, Van Houtan K (2019) The vertical distribution and biological transport of marine microplastics across the epipelagic and mesopelagic water column. Sci Rep 9(1):7843. https://doi.org/10.1038/s41598-019-44117-2

Chubarenko I, Stepanova N (2017) Microplastics in sea coastal zone: lessons learned from the Baltic amber. Environ Pollut 224:243–254. https://doi.org/10.1016/j.envpol.2017.01.085

Cincinelli A, Martellini T, Guerranti C, Scopetani C, Chelazzi D, Giarrizzo T (2019) A potpourri of microplastics in the sea surface and water column of the Mediterranean Sea. In: TrAC—trends in analytical chemistry, vol 110, pp 321–326. https://doi.org/10.1016/j.trac.2018.10.026

Egger M, Sulu-Gambari F, Lebreton L (2020) First evidence of plastic fallout fromthe Great Pacific Garbage Patch. Sci Rep 10:7495. https://doi.org/10.1038/s41598-020-64465-8

Eo S, Hong SH, Song YK, Han GM, Seo S, Shim WJ (2021) Prevalence of small high-density microplastics in the continental shelf and deep sea waters of East Asia. Water Res 200:117238. https://doi.org/10.1016/j.watres.2021.117238

Erni-Cassola G, Zadjelovic V, Gibson MI, Christie-Oleza JA (2019) Distribution of plastic polymer types in the marine environment; A meta-analysis. J Hazardous Mater 369:691–698. https://doi.org/10.1016/j.jhazmat.2019.02.067

Iorga MC, Carolina N, Lozier MS (1999) Signatures of the Mediterranean outflow from a North Atlantic climatology. 2. Diagnostic velocity fields. J Geophy Res Oceans 104(Cll):26011–26029

Iorga MC, Lozier MS (1999) Signatures of the Mediterranean outflow from a North Atlantic climatology 1. Salinity and density fields. J Geophys Res Oceans 104(C11):25985–26009. https://doi.org/10.1029/1999jc900115

Kukulka T, Proskurowski G, Morét-Ferguson S, Meyer DW, Law KL (2012) The effect of wind mixing on the vertical distribution of buoyant plastic debris. Geophys Res Lett 39(7):1–6. https://doi.org/10.1029/2012GL051116

Li D, Liu K, Li C, Peng G, Andrady AL, Wu T, Zhang Z, Wang X, Song Z, Zong C, Zhang F, Wei N, Bai M, Zhu L, Xu J, Wu H, Wang L, Chang S, Zhu W (2020) Profiling the vertical transport of microplastics in the West Pacific Ocean and the East Indian Ocean with a novel in situ filtration technique. Environ Sci Technol 54(20):12979–12988. https://doi.org/10.1021/acs.est.0c02374

Morales-Caselles C, Viejo J, Martí E, González-Fernández D, Pragnell-Raasch H, González-Gordillo JI, Montero E, Arroyo GM, Hanke G, Salvo VS, Basurko OC, Mallos N, Lebreton L, Echevarría F, van Emmerik T, Duarte CM, Gálvez JA, van Sebille E, Galgani F, Cózar A et al (2021) An inshore–offshore sorting system revealed from global classification of ocean litter. Nature Sustain 4(6):484–493. https://doi.org/10.1038/s41893-021-00720-8

Peng X, Chen M, Chen S, Dasgupta S, Xu H, Ta K, Du M, Li J, Guo Z, Bai S (2018) Microplastics contaminate the deepest part of the world's ocean. Geochemical Perspect Lett 9:1–5. https://doi.org/10.7185/geochemlet.1829

Poulain M, Mercier MJ, Brach L, Martignac M, Routaboul C, Perez E, Desjean MC, ter Halle A (2019) Small microplastics as a main contributor to plastic mass balance in the North Atlantic subtropical gyre [Research-article]. Environ Sci Technol 53. https://doi.org/10.1021/acs.est.8b05458

ten Hietbrink S, Materić D, Holzinger R, Groeskamp S, Niemann H (2025) Nanoplastic concentrations across the North Atlantic. Nature 643(8071):412–416. https://doi.org/10.1038/s41586-025-09218-1

Vega-Moreno D, Abaroa-Pérez B, Rein-loring PD, Presas-navarro C, Fraile-nuez E, Machín F (2021) Distribution and transport of microplastics in the upper 1150 m of the water column at the Eastern North Atlantic Subtropical Gyre, Canary Islands, Spain. Sci Total Environ 788:147802. https://doi.org/10.1016/j.scitotenv.2021.147802

Vega-Moreno D, Sicilia-González S, Domínguez-Hernández C, Moreira-García E, Aguiar-González B, Hernández-Borges J, Fraile-Nuez E, Machín F (2024) Exploring the origin and fate of surface and sub-surface marine microplastics in the Canary Islands region. Front Mar Sci 11:1–12. https://doi.org/10.3389/fmars.2024.1314754

Chapter 43
Meso- and Microplastics in Beach Sediments with *Posidonia oceanica*: A Preliminary Assessment from Ligurian Sea

Lisa Zanetti, Elisa Costa, Veronica Piazza, Filippo Castelli, Alessio Montarsolo, Roberta Nugnes, and Francesca Garaventa

Introduction

The ubiquitous presence of plastic debris in marine ecosystems is an increasingly recognized concern, with ecological, social, and economic implications both at local and global scales (Beaumont et al. 2019). The Mediterranean Sea is among the most affected areas, due to intense coastal urbanization, maritime activities, and its semi-enclosed configuration which limits water exchange (Tsiaras et al. 2022). The strong spatiotemporal variability in its circulation prevents the formation of stable accumulation zones, promoting continuous exchange between floating and shoreline litter (Bentaallah et al. 2024). Shorelines constitute one of the most important components of the coastal marine environment (Simantiris et al. 2025).However, they are highly vulnerable to the simultaneous influence of multiple, interacting stressors, with plastic pollution further undermining their resilience (Lincoln et al. 2022). The endemic seagrass *Posidonia oceanica*, essential for coastal protection and to provide ecosystem services, also contributes to plastic transport towards the beach environment (Sanchez-Vidal et al. 2021). Within the Mediterranean basin, the Ligurian coastline (NW Mediterranean) combines high biodiversity and protected habitats with strong pressures from urbanization, erosion, harbor infrastructures, and tourism (Cattaneo Vietti et al. 2013). In this context, assessing the extent of plastic pollution is crucial, particularly due to the limited scientific literature available for this area. This study provides a preliminary assessment of meso- (MEPs, 5–25 mm) and microplastics (MPs, < 5 mm) on two urban beaches of Genoa (Ligurian Sea),

L. Zanetti (✉) · E. Costa · V. Piazza · F. Castelli · A. Montarsolo · R. Nugnes · F. Garaventa
National Research Council, Institute for the Study of Anthropic Impact and Sustainability in the Marine Environment (CNR-IAS), Genova, Italy
e-mail: lisa.zanetti@ias.cnr.it

G. Gentile et al. (eds.), *Proceedings of the 4th International Conference on Microplastic Pollution in the Mediterranean Sea*, Springer Water,
https://doi.org/10.1007/978-3-032-23847-4_43

also evaluating their relationship with stranded *P. oceanica*. The work was carried out within the SUNRISE project, which promotes Nature-Based Solutions for coastal restoration.

Materials and Methods

Three sampling activities (April 2024, November 2024 and April 2025) were carried out on two urban beaches in the city of Genoa, Quarto (Beach A, 44°23′09″N, 9°00′24″E) and Sturla (Beach B, 44°23′33″N, 9°01′52″E). This area is characterized by different size coarse pebbly–gravelly pocket beaches (Cattaneo Vietti et al. 2013) with high touristic season during from June to September. From the structural point of view, Beach A is enclosed by multiple artificial breakwaters, Beach B is more linear, bordered by a single mole and intersected by the estuary of a small stream. Occasional accumulations of *P. oceanica* debris are present on both beaches. The superficial 5 cm of sediment, including any vegetal material, were collected from each beach, using six 50 × 50 cm quadrats located along a single transect at the lower depositional line.

In the laboratory, samples were dried (40 °C) and MEPs were extracted by sieving, while MPs were further isolated by density separation, using a Sodium Iodide (NaI) solution ($\rho = 1.9$ g/cm^3). Any vegetal material was quantified as a percentage of the total sample volume. All the presumed plastic items were characterized by shape and chemically analyzed by FTIR spettroscopy.

Differences between beaches and sampling dates were tested through univariate PERMANOVA, including *P. oceanica* quantity as a covariate, with separate analyses for MEPs and MPs abundances (items/kg).

Results

A total of 537 MEPs and 519 MPs were isolated across the three sampling dates and the two investigated beaches. MEPs and MPs mean abundances (items/kg) are reported in Table 43.1, while their distribution by beach and sampling date is shown in Fig. 43.1a (MEPs) and Fig. 43.1b (MPs).

Preliminary statistical analyses revealed no significant differences in MEPs and MPs abundances between beaches (Fig. 43.1), suggesting that neither structural differences nor the presence of the stream at site B—typically considered a potential source of plastic debris—appeared to influence plastic accumulation. Moreover, no differences were detected between pre- and post-touristic season samplings (April and November 2024). In contrast, MEPs abundance differed significantly ($p < 0.05$; Fig. 43.1a) between pre-touristic samplings of consecutive years (April 2024, April 2025), suggesting the influence of environmental drivers beyond local touristic pressure. This may reflect the high dynamicity of the Ligurian Sea (Cattaneo Vietti et al.

Table 43.1 Meso- and microplastic mean abundance values (items/kg; mean ± es), in the three sampling dates and two beaches (A and B)

Sampling date	MEPs/kg		MPs/kg	
	A	B	A	B
April 2024	1.16 ± 0.86	1.88 ± 0.74	10.51 ± 9.95	9.75 ± 2.79
November 2024	0.21 ± 0.10	7.31 ± 5.82	15.99 ± 1.96	21.51 ± 5.26
April 2025	7.38 ± 4.05	3.96 ± 2.45	28.20 ± 20.50	24.30 ± 2.45

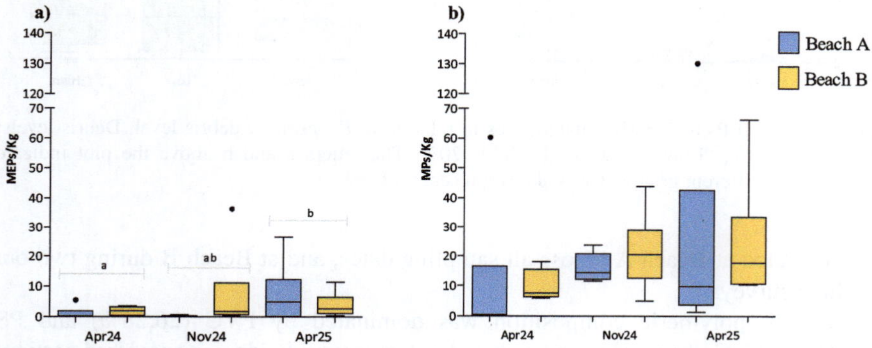

Fig. 43.1 **a** MEPs and **b** MPs abundances for beaches and sampling date. The letters a and b above the plot indicate significantly different groups. Black dots represent outliers

2013) and the peculiar morphology of pocket beaches, which may be more affected by redistribution processes, driven by hydrodynamics or stochastic events, than wider beaches. These observations are further supported by (Misic et al. 2019), who demonstrated the importance of sea conditions in influencing anthropogenic microparticles distribution along the Ligurian coast.

Results shown in Fig. 43.2a indicate a significant association ($p < 0.005$) between *P. oceanica* debris and MEPs abundance, which was higher at "high" debris level compared to "absent" level. This supports the role of seagrass detritus as a vector for plastic litter, consistent with other Mediterranean studies (Pietrelli et al. 2017; Sanchez-Vidal et al. 2021; Afeniforo et al. 2025), while no significant effect was detected for MPs (Fig. 43.2b).

Overall, of the 1056 plastic particles analyzed (MPs and MEPs), fragments were the most abundant shape (67.65%), followed by rubber-like granules (15.93%) and films (11.48%), while foams, filaments, and pellets accounted for less than 5%. Fibers were not included in this preliminary analysis. Shape distribution by size class (Fig. 43.3a for MEPs; Fig. 43.3b for MPs) confirmed fragments as the predominant category (77.94% for MEPs; 57.03% for MPs), suggesting fragmentation as the main source of plastic particles. However, a considerable proportion (32.37% of MPs and 15.41% of total particles) of black rubber-like granules (Fig. 43.3b and c)

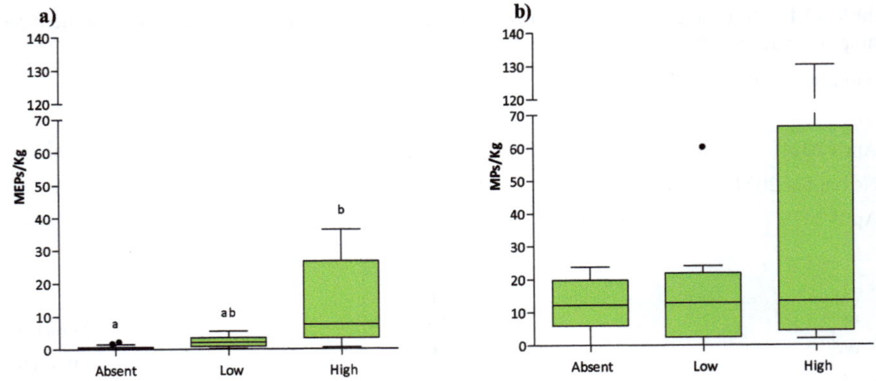

Fig. 43.2 **a** MEPs and **b** MPs abundances in relation to *P. oceanica* debris level. Debris levels: "Absent" = 0%; "Low" ≤ 20%; "High" > 20%. The letters a and b above the plot indicate significantly different groups. Black dots represent outliers

was detected at Beach A across all sampling dates, and at Beach B during two out of three surveys.

Overall, polymeric composition was dominated by PVC (26.28%) and PS (16.69%). PVC, the third most produced polymer worldwide, is considered particularly hazardous due to its poor thermal stability, the presence of toxic additives, and its rapid degradation into MPs (Campisi et al. 2025). As regards the black rubber-like granules, FTIR spectra showed absorption features consistent with EPDM (Ethylene Propylene Diene Monomer), suggesting similarities with rubber infill materials used

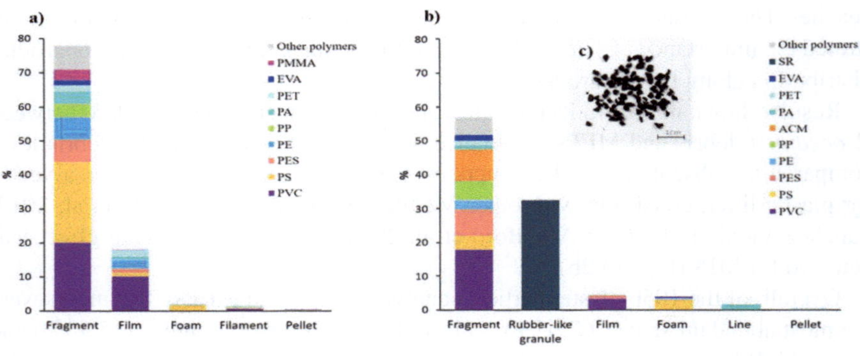

Fig. 43.3 Shape and polymer distributions of **a** MEPs and **b** MPs. "Other polymers" includes ABS (Acrylonitrile Butadiene Styrene), PVA (Polyvinyl Alcohol), PU (Polyurethane), PC (Polycarbonate), EPDM (Ethylene Propylene Diene Monomer), CA (Cellulose Acetate), Mater-B, and PLA (Polylactic Acid). Abbreviations used: PVC (Polyvinyl Chloride), PS (Polystyrene), PES (Polyester), PE (Polyethylene), PP (Polypropylene), ACM (Acrylic Rubber), PA (Polyamide), PET (Polyethylene Terephthalate), EVA (Ethylene Vinyl Acetate), PMMA (Polymethyl Methacrylate), SR (Synthetic Rubber)

in artificial turf fields (Bø et al. 2024). Since definitive polymer identification was not possible, they were conservatively classified as *Synthetic Rubber* (SR) in this preliminary study. Often derived from end of life tires, rubber infill are of particular concern, due to their widespread dispersion and the release of toxic substances such as PAHs, PFAS and heavy metals (Bø et al., 2024) Representing 38% of intentionally added MPs they are currently subject to transitional EU regulation (Zuccaro et al. 2024).

Conclusions

These preliminary findings highlight the complex interplay between environmental drivers and anthropogenic sources in influencing plastic pollution dynamics, with *P. oceanica* playing a key role. They also underline the importance of using comprehensive extraction methodologies and polymer-specific analyses to accurately assess ecological risks and supporting effective mitigation strategies.

Acknowledgements This work was carried out within the SUNRISE Project, "Le 'soluzioni basate sulla natura' per la conservazione della biodiversità marina e lo sviluppo sostenibile in Mar Ligure", coordinated by Stazione Zoologica Anton Dohrn and mainly supported by Fondazione Compagnia di San Paolo ("Simbiosi – Insieme alla natura per il futuro del Pianeta").

References

Afeniforo T et al (2025) *Posidonia oceanica* wrack intercepts plastic debris: First evaluated evidence on Maltese beaches. Reg Stud Marine Sci 90:104439

Beaumont NJ et al (2019) Global ecological, social and economic impacts of marine plastic. Mar Pollut Bull 142:189–195

Bentaallah ME et al (2024) Assessment of microplastic abundance and impact on recreational beaches along the western Algerian coastline. Mar Pollut Bull 199:116007

Bø SN et al (2024) Environmental impacts of artificial turf: a scoping review. Int J Environ Sci Technol 21:10205–10216

Campisi L et al (2025) Polyvinyl chloride (PVC), its additives, micro-plastic and human health: unresolved and emerging issues. Sci Total Environ 960:178276

Cattaneo Vietti R et al (2013) The Ligurian Sea: present status, problems and perspectives. Chem Ecol 26(S1):319–340

Lincoln S et al (2022) Marine litter and climate change: inextricably connected threats to the world's oceans. Sci Total Environ 837:155709

Misic C et al (2019) Hydrodynamic forcing and sand permeability influence the distribution of anthropogenic microparticles in beach sediment. Estuar Coast Shelf Sci 230:106429

Pietrelli L et al (2017) Pervasive plastisphere: first record of plastics in egagropiles (Posidonia spheroids). Environ Pollut 229:1032–1036

Sanchez-Vidal A et al (2021) Seagrasses provide a novel ecosystem service by trapping marine plastics. Sci Rep 11:254

Simantiris N et al (2025) A systematic review and scientometrics analysis on microplastic pollution on coastal beaches around the globe. Cont Shelf Res 286:105424
Tsiaras et al (2022) Microplastics in the Mediterranean: variability from observations and model analysis. Frontiers Marine Sci 9
Zuccaro P et al (2024) The European Union ban on microplastics includes artificial turf crumb rubber infill: other nations should follow suit. Environ Sci Technol 58(6):2591–2594

MIX
Papier aus verantwortungsvollen Quellen
Paper from responsible sources
FSC® C105338

FSC
www.fsc.org

If you have any concerns about our products,
you can contact us on
ProductSafety@springernature.com

In case Publisher is established outside the EU,
the EU authorized representative is:
Springer Nature Customer Service Center GmbH
Europaplatz 3, 69115 Heidelberg, Germany

Printed by Libri Plureos GmbH
in Hamburg, Germany